Overcoming Niagara

Overcoming Niagara

*Canals, Commerce, and Tourism in the
Niagara–Great Lakes Borderland Region,
1792–1837*

Janet Dorothy Larkin

Published by State University of New York Press, Albany

© 2018 State University of New York

All rights reserved

No part of this book may be used or reproduced in any manner whatsoever without written permission. No part of this book may be stored in a retrieval system or transmitted in any form or by any means including electronic, electrostatic, magnetic tape, mechanical, photocopying, recording, or otherwise without the prior permission in writing of the publisher.

For information, contact State University of New York Press, Albany, NY
www.sunypress.edu

Production, Eileen Nizer
Marketing, Fran Keneston

Library of Congress Cataloging-in-Publication Data

Names: Larkin, Janet Dorothy, author.
Title: Overcoming Niagara : canals, commerce, and tourism in the Niagara–Great Lakes borderland region, 1792–1837 / Janet Dorothy Larkin.
Description: Albany : State University of New York Press, [2018] | Includes bibliographical references and index. | Description based on print version record and CIP data provided by publisher; resource not viewed.
Identifiers: LCCN 2017008959 (print) | LCCN 2017012086 (ebook) | ISBN 9781438468259 (ebook) | ISBN 9781438468235 (hardcover   ) | ISBN 9781438468242 (pbk.   )
Subjects: LCSH: Canals—New York (State)—Niagara Frontier—History—18th century. | Canals—New York (State)—Niagara Frontier—History—19th century. | Niagara Frontier (N.Y.)—Commerce—History. | Niagara Frontier (N.Y.)—Description and travel. | Tourism—New York (State)—Niagara Frontier—History. | Erie Canal (N.Y.)—History. | Welland Canal (Ont.)—History. | Oswego Canal (N.Y.)—History. | Borderlands—Niagara Falls Region (N.Y. and Ont.)—History. | Borderlands—Great Lakes Region (North America)—History.
Classification: LCC HE395.N74 (ebook) | LCC HE395.N74 L37 2018 (print) | DDC 386/.409747—dc23
LC record available at https://lccn.loc.gov/2017008959

To John,
scholar, mentor, much-loved friend.

From a Drawing by W.R. Callington, Engineer, Boston, from an Actual Survey made in 1837.

A bird's-eye view of the river Niagara from Lake Erie to Lake Ontario showing the situation and extent of NAVY ISLAND, and the Towns and Villages on the banks of the river in Canada and the United States, with the situation of the Caroline Steam Boat off Schlosser.

United States
1. Town of Buffalo
2. Black Rock
3. Grand Island
4. Tonewanta Creek
5. Grand Canal
6. Hotel at the falls
7. Lewiston
8. Fort Niagara
9. Lake Ontario
10. Goat Island
11. Schlosser, the place where the Caroline was burnt

Canada
A. Lake Erie
B. Fort Erie
C. Waterloo
D. Navy Island
E. Chippewa
F. Rapids
G. Cataracts of Niagara
H. Queen's Town
I. Fort George
J. Welland Canal

Published by J. Robins, Bride Court, Fleet Street.

A Bird's-Eye View of the Niagara River, 1837. This Bird's-eye view of Niagara Falls illustrates not only the sublimity of the majestic wonder itself but also the role of the interlocking canal system in uniting Canadians and Americans of the borderland. Courtesy of the Toronto Public Library.

The Niagara, though called a river, would be more accurately described as a natural canal, of about 36 miles in length, through which the waters of Lake Erie pass northward into Lake Ontario.

—Monthly Supplement of the Penny Magazine
of the Society for the Diffusion of Useful Knowledge (1837–1838)

Contents

List of Illustrations	xi
Acknowledgments	xv
1. Introduction	1
2. The "Greatest Display in Nature": The Niagara–Great Lakes Borderland Region before the War of 1812	13
3. "As Irresistible as the 'Cataract of Niagara'": The Erie Canal and the Promises of the North American Canal Age	43
4. "A Salutary and Desirable Competition": New York State Influence in the Building of the First Welland Canal	79
5. The Oswego Canal: "One of the Great Traveling and Commercial Thoroughfares"	113
6. "The Great Wonder of the Canal"	143
7. Conclusion	177
Notes	185
Bibliography	241
Index	259

Illustrations

frontispiece **A Bird's-Eye View of the Niagara River, 1837.** This Bird's-eye view of Niagara Falls illustrates not only the sublimity of the majestic wonder itself but also the role of the interlocking canal system in uniting Canadians and Americans of the borderland. Courtesy of the Toronto Public Library.   vi

Figure 0.1 **Major Canals in the Niagara–Great Lakes Borderland Region.** Reprinted From "Freight Capacity and Utilization of the Erie and Great Lakes Canals before 1850" by Thomas F. McIlwraith. *The Journal of Economic History* (December 1976). Courtesy of Cambridge University Press.   xviii

Figure 1.1 **The Niagara Borderland.** John M. Duncan, *Travels Through Parts of the United States and Canada in 1818 and 1819*. Courtesy of the New York Public Library.   4

Figure 2.1 **Black Rock, New York.** As early as the eighteenth century, Black Rock served as a major crossing point to Canada. During the War of 1812, Black Rock became a notorious smuggling center that sent much produce to the Upper Canadian and Montreal markets. During the canal age, Black Rock also became a well-known crossing point to Canada on the infamous Underground Railroad. John W. Barber and Henry Howe. *Historical Collections of the State of New York*.   24

Figure 3.1 **Joseph Ellicott.** Holland Land Company agent Joseph Ellicott early encouraged the north-south economic orientation of the Niagara borderland by promoting internal improvements like roads and canals. Courtesy of The Buffalo History Museum. Used with permission. 47

Figure 3.2 **General Peter B. Porter.** A renowned statesmen and War of 1812 hero, Porter was also a borderland entrepreneur and friend of Upper Canada. Courtesy of The Buffalo History Museum. Used with permission. 57

Figure 3.3 **The Canal Age at Niagara.** As indicated by this illustration, the Erie Canal created new opportunities along the Niagara borderland. 74

Figure 4.1 **William Hamilton Merritt.** Statesman and Father of the Welland Canal, William Merritt was often affectionately dubbed the "De Witt Clinton of Canada." J. P. Merritt, *Biography of the Honorable W. H. Merritt*. Image courtesy of the University of Alberta. 81

Figure 4.2 **De Witt Clinton.** The leading force behind the Erie Canal, Clinton's untimely death in 1828 was mourned on both sides of the Canada–United States border. Courtesy of New York Public Library Digital Collection. 85

Figure 4.3 **John B. Yates.** American investor John B. Yates held the largest number of private shares in the Canadian Welland Canal Company and was a keen champion of internal improvements on both sides of the Canada–United States border. Without Yates's financial and moral support, work on the Welland Canal would have been halted more than once. Courtesy of the Village of Chittenango, New York. 87

Figure 4.4 **Buffalo Journal Advertisement.** American newspapers like the *Buffalo Journal* regularly advertised employment and other related opportunities on Canada's Welland Canal. Courtesy of the *Buffalo Journal*. 95

Figure 4.5 **The Welland and Erie Canals within the Niagara Borderland.** From John N. Jackson, *The Mighty Niagara: One River, Two Frontiers*. Used with permission. 102

Figure 5.1 **A Rare View of the Oswego Canal during the 1830s.** The Oswego Canal, especially in conjunction with the Welland Canal, facilitated closer transportation and commercial ties with neighboring Canada, while also providing an indispensable link in the Northern Tour. John W. Barber and Henry Lowe, *Historical Collections of the State of New York*. Image courtesy of the Library of Congress. 115

Figure 5.2 **Gerrit Smith.** Smith was not only a leading American abolitionist and reformer, but the largest investor in the Oswego Canal and friend of open and free trade across the Canada–United States border. Octavius Brooks Frothingham, *Gerrit Smith: A Biography*. Image courtesy of Cornell University. 118

Figure 6.1 **Dwight's Map of the Erie and Welland Canal System along the Niagara Frontier.** As indicated by Dwight's map, man-made wonders like the Erie and Welland Canals were central to the visitor's experience at Niagara during the nineteenth-century age of transportation and innovation. Theodore Dwight, *The Northern Traveller*. Image courtesy of the Library of Congress. 144

Figure 6.2 **Salina Salt Works.** During the nineteenth century, Salina became one of the nation's most important salt manufacturing industries to send its profitable commerce to both American and Canadian markets. Revenue from these sales helped finance the New York canal system. Salina was also a popular stopover on the fashionable Northern Tour. John W. Barber and Henry Howe, *Historical Collections of the State of New York*. Image courtesy of the Library of Congress. 149

Figure 6.3 **Travel Options on the Northern Tour.** Part of the excitement of the Northern Tour was to diversify the travel itinerary and opt for the latest mode of land and water craft that the canal age had to offer. Courtesy of the *Black Rock Gazette*. 156

Figure 6.4	**Lewiston, NY, and Queenston, Upper Canada, Ferry Landings.** As a result of canal and transportation innovations at Niagara, ferries and steamboats regularly plied between the Canadian and American sides. John W. Barber and Henry Howe, *Historical Collections of the State of New York*. Image courtesy of the Library of Congress.	158
Figure 6.5	***Michigan* Broadside.** Courtesy of the Buffalo History Museum.	163
Figure 6.6	**Entrance to Lockport on the Erie Canal.** Travelers to Niagara were as fascinated by innovations in canals and other internal improvements as they were by the sublimity of the Falls. The Miriam and Ira D. Wallach Division of Art, Prints, and Photographs: Print Collection, the New York Public Library. "Entrance to the Harbor, Lockport." Courtesy of the New York Public Library Digital Collections.	171
Figure 7.1	**The *Vandalia*.** One of the first steam-driven screw-propelled ships in North America. The *Vandalia* was built in Oswego, New York, specifically for trade on the Welland Canal. Courtesy of the Buffalo History Museum.	178

Acknowledgments

One of my fondest memories of Niagara Falls is connected to my mother. When I was a kid growing up in southern Ontario, I remember her waking me on a school day and asking: "Do you feel like going to Niagara Falls today?" Naturally, I responded with a resounding yes. However, my enthusiasm for visiting Niagara that day had as much to do with the opportunity to skip school as it did with seeing the Falls. Ironically, years later it was my passion for history that drew me back to Niagara. I mark the beginning of my fascination with this remarkable wonder to that whimsical journey I took with my mother many moons ago.

So many librarians, archivists, and staff from both sides of the Canada–United States border made possible this book's completion. On the American side, I would like to thank the staff at the Buffalo History Museum; the Buffalo and Erie County Public Library (Grosvenor Room); the Niagara County Historical Society (Lockport); the Niagara Falls Library; the University of Buffalo (Lockwood Library); the Library Annex (Cornell University); the New York State Archives; the New York Public Library; the New York Historical Society; and the Onondaga Historical Association. On the Canadian side, special thanks to Edie Williams and the staff at Brock University Archives; the St. Catharines Museum; the Welland Canal Centre Lock 3; the Ontario Historical Society; the Toronto Public Library; the Niagara Historical Society (Niagara-on-the-Lake); and the Niagara Falls Public Library. Their generous time and assistance is greatly appreciated.

I would also like to express my gratitude to the historians and staff who brought my attention to, or permitted me to use, several illustrations in this book. I am especially grateful to Charles Albee (Historian for the village of Chittenango, New York); Cynthia Van Ness (Director of Library and Archives at the Buffalo History Museum); Shane Stephenson (Library Technician at the Buffalo History Museum) and Amy Miller (Assistant

Librarian at the Buffalo History Museum). The staff at the St. Catharines Museum also deserves special thanks for their assistance with images and illustrations.

One also incurs many personal and intellectual debts when writing a book. I am extremely indebted to Michael Frisch who read the entire manuscript and offered kind criticism and moral support along the way. Thomas McIlwraith also provided warm encouragement and collegial support, which made the challenges of researching and writing that much more enjoyable. Ellen Litwicki and David Kinkela read earlier parts of the book and encouraged me to move forward. Douglas McCalla graciously answered my questions about commercial matters tied to the Great Lakes region. Special thanks are also due Arthur Bowler, Patrick McGreevy, Andrew Holman, Carl Benn, Walter Lewis, Barbara Carmichael, John Staples, Jennifer Hildebrand, and Mary Beth Sievens.

During my doctoral studies at the University of Buffalo, I was fortunate to make many lasting friendships. In particular, I wish to acknowledge Thomas Grace, Robert Caputi, Paul Rodell, David Gerber, and Carolyn Korsmeyer. I have benefited in untold ways from their intellectual and personal support and encouragement.

Sadly, several colleagues and friends did not live to see the completion of this book. However I will be forever grateful to the late Richard Ellis, the late Roberta Styran, the late John Jackson, the late Frederick Drake, and the late Nicholas Cushner, all of whom took a keen interest in my work.

Much time and assistance was provided by the staff at SUNY Press. Particular thanks to Amanda Lanne-Camilli, Chelsea Miller, Eileen Nizer, Fran Keneston, and Sharon Green. This book also benefitted from outside readers who kindly read the manuscript and offered invaluable suggestions and insights.

Thanks also to the many historians and scholars, past and present, in the field of canals, borderlands, regional history, and Canadian-American relations in general who provided the inspiration and background information for the writing of this book.

The challenges of writing this book were made bearable by the support of close friends. Special thanks to Kim Madden, Elizabeth Cleary, Virginia Elliott, Cindy Kelly, Judy Larkin, Charleen Capwell, Janice Cushner, Skip Richardson, and Harvey Pines.

I am also fortunate to have a supportive and loving family. My beloved mother Dorothy Rose Baglier instilled in me the importance of obtaining a higher education and the belief that any and all goals are attainable if you apply yourself. I would also like to thank my brother Robert Baglier who during my undergraduate years guided me through the halls of academe for which I am forever grateful. Coincidently, my brother Gordon Baglier served as an engineer on the present Welland Canal, thereby following in the footsteps of the many nineteenth-century engineers and contractors who made possible this vital waterway. My life is also fuller because of my nephew Samuel Baglieri, my sister-in-law, Cathy Proctor, and Patricia Arnburg. Thanks also to the entire Larkin clan who have provided me with a warm and loving extended family.

Finally, I wish to express my deepest gratitude to my husband and devoted companion of more than twenty-five years, John Alan Larkin. His wise counsel and unwavering support allowed me to complete this work.

As is customary to state, any errors or oversights in this book are my own.

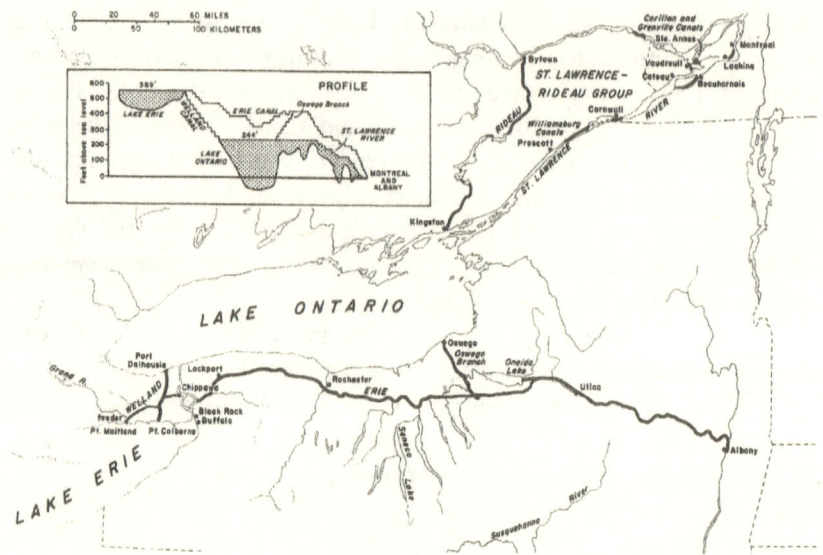

Figure 0.1. **Major Canals in the Niagara–Great Lakes Borderland Region.** Reprinted From "Freight Capacity and Utilization of the Erie and Great Lakes Canals before 1850" by Thomas F. McIlwraith. *The Journal of Economic History* (December 1976). Courtesy of Cambridge University Press.

1

Introduction

> There is yet another and very important channel. . . . I mean the Welland Canal, cut across the Isthmus of Niagara in Upper Canada, which by uniting Lake Erie with Lake Ontario, affords a communication between the western lakes and the seas, either through the St. Lawrence, or by the Oswego Canal to Syracuse, and thence by the Grand Canal to the port of New York.
>
> —Captain Basil Hall, *Travels in North America*[1]

Captain Basil Hall was a renowned British naval officer and classic travel writer who visited North America in 1827 and 1828. During this time, Hall journeyed extensively throughout the continent, carefully recording his impressions of the advances in land and water transportation that heralded the ensuing canal age. As a captain of the Royal Navy who had spent nearly half of his life in service to the British Empire, Hall was fascinated with developments in transportation in the United States and Canada, and their impact on the age-old question of competition and commercial rivalry. After several weeks inspecting the Erie and Welland Canals, Hall offered some surprising observations on these rival transportation systems in North America. "At first sight," he wrote, "it may seem that the Welland Canal, by offering superior advantages, will draw away from New York a portion of the rich produce of the state of Ohio, of Upper Canada, and of the other boundless fertile regions which form the shores of the higher lakes, yet there seems little doubt that the actual production of materials requiring transport will increase still faster . . . and that ere long additional canals, besides these two, will be found necessary." Hall concluded that "the upper countries alluded to will derive considerable advantages from having a free choice of markets, as they may now proceed either to New York by the Erie Canal, or by the Welland Canal, down the river St. Lawrence, according as the market of New York or that of Montreal shall happen to be the most favourable, or the means of transport cheapest."[2]

Hall's emphasis on the complementarities between the New York and Welland canal systems contrasts with much scholarly and popular writing about these canals and the larger stories around them—Canadian-American commercial competition, national rivalry, and westward expansion and market development generally. Whether in historiography or popular understanding, the story of the competing Canadian-American canal systems, and their broader narratives, seems a pivotal, settled, and almost legendary story with a clear message. The classical account begins with New York City-led economic expansion and America's vulnerable military position along the New York frontier during the War of 1812. Seeking a leg up in the competition for the lucrative trade of the Midwest, and fearing another war with Great Britain and Canada, New York State built an interior canal to Lake Erie that bypassed British territory altogether. An alternative "Lake Ontario route" called for a canal around Niagara Falls that would have connected Lakes Erie and Ontario, and then linked the Oswego to the Mohawk and Hudson Rivers. But it was feared that once trade reached Lake Ontario, it would be lost to the St. Lawrence and Montreal market. By opting for the direct route between Lake Erie and the Hudson, Niagara Falls was deliberately left as a major barrier between Montreal and the interior.[3] However, according to the same settled story, Canada swiftly responded to the Erie Canal by building the Welland ship canal between Lakes Ontario and Erie—a vital component in the St. Lawrence–Great Lakes water system that allowed Canadians to compete more effectively for the western trade, while also creating a market between the disparate provinces.[4] Again, the United States went on the offensive by undertaking massive improvements to the New York canal system, but the introduction of the railroad, combined with the Panic of 1837 and continuing improvements to the St. Lawrence–Welland seaway, diminished the Erie Canal's importance. Much canal scholarship continues to echo this conventional account.[5]

This classic understanding of Canadian-American national rivalry and competition has been seemingly confirmed by twentieth-century developments. The successful Welland Canal became an integral part of the St. Lawrence Seaway in 1954, but the Erie Canal sank into oblivion. Buffalo's once vibrant position as a leading commercial port on Lake Erie was made obsolete following the enlargement and improvement of the Welland system.[6] Even today, as New York State focuses on revitalizing the Erie Canal through tourism and recreation, the Welland Canal continues to serve as a viable part of the seaway, allowing large lake vessels and supertankers to navigate in and out of the continent.

Of all the borderland regions in North America, Niagara's location has been central in shaping the conventional story of conflict and rivalry between the Canadian and American transportation systems and their broader narratives. The Niagara River and Falls forms a natural barrier between the two peoples, and since 1783 Canadians and Americans found themselves on opposite sides of an international border that created the newly established British province of Upper Canada and the American Republic (fig. 1.1). As the *water gateway* to the West, the Niagara River and Falls held strategic value for both countries. As one historian observed "in the long-range commercial strategy of New York State" the Niagara barrier played a vital role in the Erie Canal's building. The threat posed to national security during the War of 1812, and the questionable loyalty of many upstate New Yorkers who took advantage of Canadian markets during the protracted conflict, helped persuade the New York legislators that the Niagara River must remain a barrier because "they could circumvent it by means of the Erie Canal, whereas Montreal could not."[7] In Canada, the Welland Canal's location has similarly been viewed as a direct response to America's propinquity along the Niagara frontier during the 1812 imbroglio. "The memory of this conflict," wrote one Canadian source, ruled against a canal at Niagara. Instead, in this same view, the Welland was deliberately built some considerable distance from the American frontier.[8] Canada's Niagara peninsula, which pointed "like a spear at the heart of the American union," became a symbol for the contested canal age in this region.[9]

It was also at Niagara that North America's three most important canals—the Erie, Oswego, and Welland—connected and converged, making travel more accessible and opening the region to commerce, tourism, and improvement in general on a grand scale. Yet, the Niagara tourist industry, like the canals, has similarly been described in terms of competing national ideologies and interests. The story of this attraction, and tourism generally, has been seen through the lens of the natural wonder of the Falls itself, and the natural border it marks between the United States and Canada. For Elizabeth McKinsey, the power manifested in Niagara Falls becomes an icon of American prowess and patriotism. Patrick McGreevy, who has

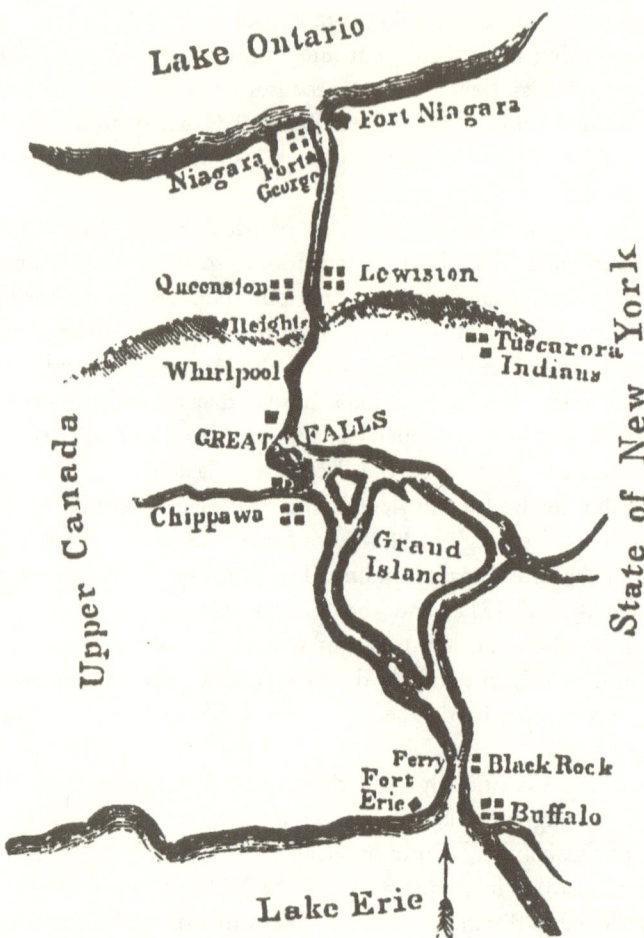

Figure 1.1. **The Niagara Borderland.** John M. Duncan, *Travels Through Parts of the United States and Canada in 1818 and 1819.* Courtesy of the New York Public Library.

written extensively on Niagara Falls, argues in his thought-provoking article "The End of America: The Beginning of Canada"(1988) that the Falls have different meanings for the United States and Canada which can only be understood "in relation to two very different ideologies of nationalism."[10] Just as nation-centered history has shaped much writing on the canal age, scholars have discussed the tourist industry as distinctive and separate United States and Canadian stories.

This book tells a different story—one that examines the canal age from a borderland and transnational perspective. Despite the existence of the international boundary line and the ongoing geopolitical contest between the United States and Great Britain for control of the continental interior, Canadians and Americans embarked on a remarkable series of canal projects and other cross-border improvements driven not simply by continental rivalries and routings, but by the life, commerce, and connectivity of the Niagara region. Uninhibited by the international border, residents in upstate New York and Upper Canada built roads, bridges, postal and ferry services, and water transport systems that allowed for easier access to and around the Falls, while also creating an interlocking, interconnected transportation network in the Niagara–Great Lakes Basin. Personal and familial ties forged by the late loyalists, the existence of an early cross-lake trade, and the potential to develop commerce and tourism in the region contributed to the porosity of the border. By the War of 1812, and even during and after, lucrative commercial and cultural connections continued to be forged across the border.[11] This book's title, *Overcoming Niagara*, refers to the many ways in which Canadians and Americans mutually sought to overcome natural and artificial barriers, building an integrated, interlocking canal system that strengthened the borderland economy while also propelling westward expansion, market development, tourism, and progress generally.

A major premise of this book is that north-south transportation and communication linkages in the Niagara–Great Lakes region suggest little relationship between the international boundary line and the broader geopolitical interests of the United States and Great Britain.[12] Nowhere was this more evident than during the building of the Erie Canal. Despite the broader narrative of Canadian-American commercial competition and rivalry, Upper Canada contributed to, and benefited from the Erie Canal in untold ways. Indeed, the historic debate over the Lake Ontario route's challenge to the Erie Canal's route revolved far less around a United States–Canada or New York–Montreal rivalry, than around the claim that the lake route might strengthen commercial ties with Canada while also sending substantial trade to the New York market. Many American merchants and shippers were awakened to the profitable, if illegal trade with Canada during the embargo and War of 1812 and in the postwar era became keen champions of the Lake Ontario route and closer Canadian connections. Even Erie Canal founder De Witt Clinton, who inspected both routes in 1810 as Canal Commissioner, spoke favorably of the Lake Ontario channel's capacity to bring economic and personal benefits to both sides of the border. No matter where a person stood on the Erie Canal, or on which side of the boundary they resided, the advantages of maintaining close ties between the neighboring countries were rarely disputed.

While conventional histories emphasize the Erie Canal's role in creating a canal boom throughout the United States, little attention is given to the American channel's influence in sparking enthusiasm, and making possible, the construction of the Welland Canal in neighboring Canada.[13] The Welland Canal projectors looked to the United States for support in their undertaking, knowing that the Erie Canal was almost completed, and that surplus laborers, engineers, contractors, technology, and capital would be available to assist in the upper province. Considered in its North American context, the Welland Canal (which unlike the Erie barge canal promised to serve large Great Lakes vessels) was welcomed as alleviating some of the early frustrations and delays being reported on the American channel, while at the same time opening a potentially more competitive and efficient means of moving American goods in and out of the interior. Conversely, the Welland Canal provided a major conduit through which Canadian goods could be transshipped to New York, thanks to the Erie and soon to open Oswego Canal. Whether destined for the Montreal or New York markets, the Welland Canal promised to facilitate trade on both sides of the border. It was this compatibility and interconnectivity that Captain Basil Hall spoke of in the quote opening this introduction where he described the New York and Welland Canal system in terms of a "a generous and legitimate rivalry."

To focus on a uniquely *Canadian* or *American* canal age obscures the complex and nuanced history of the transportation era in the Niagara–Great Lakes region. The building of the Canadian and American canal systems was more about promoting cross-border linkages and associations than it was about conflict and rivalry. Nothing underscores the persuasiveness of the borderlands' approach more than the story of the Oswego Canal. A major commercial link in the chain of inland navigation between the United States and Canada, the Oswego Canal was more than a feeder or *lateral* extension of New York's grand Erie Canal.[14] As often described in nation-centered history, the Oswego Canal was designed to offset improvements on the St. Lawrence–Welland connector that threatened to take trade from the American side. By opening a channel linking the Oswego port on Lake Ontario to the Erie Canal at Syracuse, trade would be re-routed back into the Erie Canal system, and America's bid for the West restored.[15] However, when viewed through the borderland and transnational lens, the Oswego Canal looks very different—not only was it a cord binding east and west, but an international waterway funneling goods through the Welland Canal in transit to both Canadian and American markets. A contemporary American newspaper celebrated the Oswego's development announcing: "Another triumph of human ingenuity and wisdom is achieved. The hitherto insurmountable barrier of the Niagara is overcome and the waters of the Erie may now mingle with those of Ontario, bearing upon their bosoms the bounties of civilization and the gifts of the arts."[16] Meanwhile, across the border in Upper Canada, the Oswego Canal was being heralded as an important inland improvement, permitting American traffic from Lake Ontario and Erie to flow into the Welland Canal, bringing added utility and business to the Canadian system. In the following years, the newly fashioned Oswego-Welland line served as a major thoroughfare for inland merchants, travelers, and commerce on both sides of the international border.[17] As will be demonstrated in detail in chapter 5, the Oswego's building was not simply some strategic move in the larger imperial contest for military and political dominance of the interior so much as a complementary and logical component of the integrated Canada–United States transportation network emerging in this cross-border region.[18]

Overcoming Niagara tells the story of how an emerging Canadian-American canal system promoted commerce, market development, tourism, and progress generally in the Niagara–Great Lakes basin. Though it deals with commerce and trade, it is not fundamentally an economic study so much as a broader history of development that traces the many linkages that transcended the border and helped shape a transnational region. While *Overcoming Niagara* shows that a significant cross-border trade occurred in this porous region, trade data on which economic studies depend was not systematically recorded for the period under review. Scholars note for example, that in Upper Canada "no systematic records of exports or imports were kept prior to 1849" and similar problems relate to trade data on the American side.[19] However, as more than one scholar attests, export trades and their related industries do not necessarily or only reflect local regional development and activity.[20] Smuggling for example, while not quantifiable, was pervasive and important to the economic health of the borderland community. The existence of "a large clandestine trade in butter and eggs across the Niagara River into Buffalo," and "a notoriously large illicit trade in horses and cattle" indicate a substantial and regular flow of trade back and forth across the border.[21] As discussed more thoroughly throughout the following pages, during the period of the embargoes, smuggling was such a common practice that "smuggling agents" could be hired to facilitate the illicit trade.[22] In the wake of the transportation revolution, smugglers regularly ran goods to and from the New York and Welland canal systems.[23] While not recordable in terms of volume or tonnage, smuggling was commonplace, and constituted an important measure of cross-border trade.

In addition to smuggling, population movements and settlement patterns provide convincing evidence of cross-border economic activity in the Niagara–Great Lakes borderland region. In his study of the Genesee Valley before the Erie Canal, Neil McNall estimated that sales to "migrants, dependent settlers, and new comers to Upper Canada exceeded the marketing of goods to Albany and down the Susquehanna and St. Lawrence Rivers." Though primarily interested in Canadian development, Douglas McCalla noted significant cross-border trades between Upper Canada and the adjacent American states for the same period. Following the War of 1812 and the onset of the canal age, a sustained flow of capital, goods, livestock, migrant laborers, technology, and information crisscrossed the border providing further evidence of transnational economic linkages. Local residents engaged in commerce and cross-lake trade, schooners and steamers plied the lakes funneling products in and out of both countries' canal systems, and "judging by typical cargo manifests, they also carried goods headed for local or regional destinations on both shores."[24] Both countries leading canal visionaries and promoters invested heavily in the idea that economic growth and prosperity would follow north-south transportation and communication lines.

The same canal sponsors who saw American and Canadian commercial developments as mutually compatible also saw great social and recreational significance in the integrated canal system. Because many canal leaders in upstate New York and Upper Canada were affiliated with the benevolent reform movement that swept through North America during the 1820s and 1830s, social and moral improvement found its way to the hundreds of canal- and boatmen who labored on the international waterways and lakes. Scholars have long recognized the importance of the Erie Canal in fermenting social activism along its eastern and westerly paths, but virtually no attention has been given to the north-south direction of these benevolence societies that made their way through the interlocking New York and Welland canal system.[25] The cross-border nature of reform in this region indicates how nation-centered history has obscured the transnational orientation of the reform era that brought improvement to both sides of the boundary line.

Cleaning up North America's inland canals, rivers, and lakes also aided commerce and tourism at Niagara during the second quarter of the nineteenth century. Canadian and American canal promoters who profited from the commercial waterways and their moral improvement gave equal attention to the complementary New York and Welland canal systems in promoting tourism in the region, and as a tourist lure themselves. While much scholarly attention has focused on the Erie Canal's importance in creating a tourist boom at Niagara, little focus has been given to the Welland, or for that matter the Oswego Canal, in popularizing a more encompassing Northern Tour that brought swarms of visitors to the region. This perspective on the relation of the canals to the Falls will, in chapter 6, suggest another dimension of *Overcoming Niagara*. While scholars focus on the tourists' contemplation of nature's sublime and picturesque as they approached the Niagara region on an Erie Canal packet, I concentrate on how canals and other internal improvements were attractions *in their own right*, creating as much wonder and awe as did the majestic Cataract itself.[26] Indeed, sources suggest that these man-made wonders were as fascinating to visitors as the natural wonder of the Falls, and a big part of the tourist experience. Now, in addition to observing the landscape or popular battlefields and ruins in the wake of the War of 1812, themed tours of the Erie, Oswego, and Welland Canals were added to the tourist's itinerary. Travel and emigrant guides, newspapers, and hotel advertisements all sponsored the trans-border tourist agenda despite later writing framing Canadian and American tourist industries as separate and competing experiences. By emphasizing cooperation and connectivity, rather than conflict and rivalry, it becomes clear how the Canada–United States transportation revolution overcame natural and artificial barriers to propel economic development, and even shaped the nature of tourism and progress in general.

This study focuses centrally on the concerns, interests, and motivations of local politicians, canal leaders, businessmen, and developers—all of whom were crucially involved in the inland canal systems. While government leaders and investors in the more distant metropolises may have been concerned about losing out to the competing St. Lawrence and New York commercial systems, in the Niagara–Great Lakes Basin it was the local shippers, merchants, forwarders, and businessmen who saw benefits in the alternative routes and markets of an integrated, interconnected canal system—regardless of which flag their produce sailed under. Equally important to this study are the thousands of migrant laborers who worked on the interlocking, interconnected canal system, and the concomitant growth of the reform spirit that found its way to both countries through the Erie, Oswego, and Welland lines. Because commerce and transportation were largely male domains, this study necessarily focuses on their experiences. However, where sources permit, this study acknowledges women, Native peoples, and slaves whose stories are revealed through the efforts of reformers, as well as local canal leaders and promoters who supported progress more generally in the Niagara–Great Lakes borderland region.[27]

In terms of my approach to this subject, *Overcoming Niagara* follows the story of the canal age in the Niagara–Great Lakes Basin from its early inception in 1792 until the end of the first major phase of canal construction in North America in 1837. Chapter 2 analyzes the development of cross-border transportation and tourism in the vital Niagara region in the years leading up to the War of 1812. Isolated from their respective metropolises, residents from New York and Upper Canada promoted north-south transportation and communication lines resulting in increased commercial, recreational, and personal opportunities across their shared border. As the North American Canal Era unfolded during the first half of the nineteenth century, transnational linkages would develop more tangibly as both countries undertook the building of a vast integrated, interconnected transportation system that promoted westward settlement, expansion, and improvement generally.

The next three chapters analyze North America's most important canals—the Erie, Oswego, and Welland—as cross-border commercial and cultural linkages along the northern borderland. The importance of maintaining close commercial, recreational, and personal ties with Canada as witnessed in the historic debate over the Erie Canal's route is the theme of chapter 3. Chapter 4 demonstrates that the Welland Canal was the logical next step in bringing cross-border commercial and recreational development and expansion to the Niagara–Great Lakes Basin, not a reaction to the threat of the Erie Canal. Chapter 5 continues the theme of Canadian-American cross-border ties as most tangibly embodied in the story of the Oswego Canal as a critical link to Upper Canada through the Welland Canal, bringing business and progress to New York State, while also facilitating greater commercial and recreational ties with the neighboring province. This chapter also focuses on the wave of Canadian and American reformers who zealously carried their message of spiritual and moral improvement through the interconnected New York and Welland canal systems.

The final chapter analyzes the connection between the canal age and tourism at Niagara. After showing how canals and other internal improvements at Niagara gave new shape and meaning to the Northern Tour as a distinctly transnational experience, the chapter culminates in a unique reading of a bizarre tourist spectacle. In 1827, many thousands of spectators were drawn to Niagara to watch the schooner *Michigan*, with a cargo of live animals on board, go crashing over the stupendous Falls. This bizarre event has long fascinated historians of tourism and spectacle, but on closer inspection, the chapter shows that the whole episode can be read as a surprisingly explicit and even literal comment on the changing commercial and cultural landscape of the region during the height of the North American canal age.

Any study that emphasizes a borderlands approach to the canal age must address the sensitive question of aggressive American expansionism.[28] In contrast to conventional history seeing the Erie and its tributary canals as part of a larger political and commercial plan to secure dominance of the continental interior and Canada, I will consider how common interests in promoting transportation, trade, and tourism, and improvement generally had the effect of constraining and reshaping ideas of expansionism in the Niagara–Great Lakes Basin.[29] As one scholar observed, "The northern boundary between the secessionist republic and the continuing British Empire" was not a cause of concern or focus of American territorial expansion. Indeed, had the United States coveted the region during the War of 1812, Americans "most directly bound into the hydrography" of the St. Lawrence and broader Great Lakes would have strongly opposed it. Furthermore, the same author concluded, the completion of the Erie, Champlain, and Oswego Canals after 1815 removed the problem of the British stranglehold on the St. Lawrence, thereby defusing "any potential American pressure for redefinition of boundaries or controls relating to this great natural—but naturally limited—trafficway."[30] In a similar vein, another scholar wrote: "Despite the integrative logic of the transportation system and the jingoistic words and covetous glances that some American politicians and businessmen cast toward Canada, it might be quipped that America's 'republic of farmers' did not take the land that lay to their north because, in the main, America's famers—and their commercial capitalist elites—did not wish to take it (and certainly not at the risk of war with Britain)."[31] *Overcoming Niagara* considers how the complementary and cooperative nature of the Canada–United States transportation revolution muffled the more distant expansionist tendencies in the porous Niagara–Great Lakes region.

In *Permeable Border*, the aptly titled study of the Great Lakes Basin between 1650 and 1990, co-author John Bukowczyk writes that "the transnational approach has largely been a missing link in understanding the Canada–United States relationship."[32] As a counterpoint to the familiar nation-centered narrative, *Overcoming Niagara* explores the commercial and cultural linkages of the two peoples as they mutually created an interconnected, interlocking system of navigation in the Great Lakes Basin—a system that overlapped, and at times conflicted, with both countries broader visions of transportation. While government officials and national/colonial leaders dreamed of transportation projects to bolster their commercial and military defenses and bind the distant frontier settlements to their respective metropolises, many merchants, politicians, and business leaders in the Niagara–Great Lakes region were making calculations based on their local interests and needs rather than ideology, political loyalties, or national ambitions. While not immune to the larger continental forces, this study focuses on the local and regional cast of players whose role in promoting cross-border commerce, tourism, expansion, and progress generally redefined and reshaped the meaning and outcome of the North American canal age in this region. The following pages tell the story of how frontier people imagined and created canals and other internal improvements that reshaped the economic, social, and cultural landscape, thereby overcoming Niagara and turning the barrier of the Falls into a symbol of regional cross-border interconnectedness, commonality, and development.

2

The "Greatest Display in Nature"

The Niagara–Great Lakes Borderland Region before the War of 1812

> The banks of this great outlet of the lakes are under high cultivation and seemingly alive with farms, stocks and herds, while industry is plying the oar and lifting the sail, on the bosom of its waters. This wonderful combination of such immense objects opening at once upon the view, while the tremendous roar of Niagara Falls is still thundering in your ear . . . is calculated to produce emotions which no pen can describe.
>
> —*The Emerald*, 1811[1]

In the wake of the American Revolution and the 1783 peace treaty creating the border between British North America and the United States, settlers from upstate New York and Upper Canada began the process of building their new societies and living together in peace. Toward this end, local merchants, business leaders, developers, and entrepreneurs from both sides of the Niagara border undertook the construction of roads, bridges, ferries, and improved water transport systems that would encourage easier access to and around Niagara Falls, while also promoting social, commercial, and recreational integration and unity between the neighboring countries. Of particular importance to the advancement of the borderland region was the need to overcome the navigational impediments of the Niagara River and Falls that united the Upper Great Lakes: Superior, Huron, Michigan, and Erie, with Lake Ontario and the tributary waters below. The utility of this vast water system was neither lost on traveler, trader, canal enthusiast, nor local entrepreneur, who early envisioned a ship canal around Niagara Falls. In their efforts to unlock this key navigational thoroughfare, Upper Canadian and American canal leaders and supporters helped forge social, commercial, and recreational ties across their shared boundary line. In the process, a transnational region was constructed independent of the border that now divided them.

Situated deep within the continental interior, and isolated from their respective nations' capitals, settlers in the Niagara–Great Lakes region naturally looked across the border for social, cultural, and commercial sustenance and prosperity. The settlement of the region began with the swell of settlers into western New York and Upper Canada following the American Revolution. Next to the native population and a handful of surveyors, merchants, and businessmen, most of the region's settlers were farmers who came in search of land and opportunities. While these land seekers simultaneously peopled the British province and the neighboring American territory, a large influx of *late loyalists*[2] from the United States into Upper Canada illuminated the permeability of the international border or what one scholar called "a single region of settlement."[3] An abundance of cheap land, proximity to family and friends in the United States,[4] and the desire to better one's economic position encouraged waves of American emigrants, mainly from New York, but some also from New Jersey, Pennsylvania, and Vermont, to come to Upper Canada, the largest number finding their way to the province between 1792 and 1812.[5] The influence of the late loyalists on Upper Canada's early development was evident—as a result of the influx of American settlers, the population of Upper Canada rose from six thousand in 1785 to fourteen thousand in 1791 and sixty thousand some ten years later.[6] By the eve of the War of 1812, the late loyalists were the majority population in Upper Canada. As one authority on the subject observes, "there was little to distinguish the rural settlements in the backwoods of Upper Canada from those on the frontiers of New York or the New England States." Moreover, "Although Upper Canada was politically, and for many, emotionally, a British colony, no resident before or indeed after the War of 1812 could have realistically denied that it was also a North American community."[7] Interestingly, loyalist-allied Iroquois also moved from New York to Upper Canada after the American Revolution. Many of these tribe members settled along the Grand River west of Niagara, only to find their lives again disrupted by the building of the Welland Canal and the larger transportation and market revolution engulfing the continent.[8]

Geographical proximity and shared experiences and interests between the settlers of upstate New York and Upper Canada promoted cross-border integration in this region. Despite some British concerns that the late loyalists were agents of American expansionism, many Upper Canadian politicians, merchants, and businessmen welcomed the Yankee newcomers whose skills and stamina would promote the frontier economy. Even First Lt. Governor John Graves Simcoe of Newark, Upper Canada, despite his pro-British stance, welcomed the large influx of American emigrants who did not let the international border get in the way of their personal and economic ambitions in the neighboring province. Meanwhile, Upper Canadian mercantile giant Robert Hamilton, who fought and suffered on the British side during the Revolution and now dominated the portaging of goods on the Canadian side, preferred American over European settlers who knew little of farming requirements in the raw North American frontier. In Hamilton's opinion, settlers in sufficient numbers were "Better disposed and fitter for our purpose from the United States."[9] According to Hamilton's chief biographer, "Hamilton when speaking on the subject of American development" emphasized the realities of the North American environment: "Americans adapted to local conditions more rapidly than Europeans and made better farmers." Accordingly, "Hamilton emphasized the ties of British subjects and Americans rather than their differences."[10] As to the new emigrants' political ambitions and loyalties, most British and Upper Canadian leaders conceded that the late loyalists had no "hostile or treacherous" reasons for coming but to "better their circumstances, by acquiring land upon easy terms."[11] Historian Reginald Stuart agrees that the American settlers were "frontier migrants seeking to better themselves or to find refuge. Liberal land terms and an open border was the attraction, not ideological fervor against loyalists or monarchism."[12] The arrival of the late loyalists from New York, notes immigration historian Lee Hansen, was "a thoroughly nonpolitical migration to the north and west" by people who came to the upper province "with no thought of the boundary which the diplomats had laid down in 1783."[13]

As more people flooded the region, and the infant settlements along the Niagara borderland began to take shape, the need to make Niagara Falls more accessible between the Great Lakes and the Atlantic became obvious to transportation and communication leaders on both sides of the boundary line. Indeed, the state of transportation and communications in the early years of settlement were rudimentary at best. Travelers coming from the east had two basic routes to choose from. The first began in New York

City, up the Hudson to Albany, across the Mohawk River to Oswego and Lake Ontario, and west to Niagara. Alternatively, travelers might begin their journey at Montreal, follow the St. Lawrence River to Kingston, and cross the vast expanse of Lake Ontario to the mouth of the Niagara River. Even with all of the improvements in transportation and technology that came out of the nineteenth-century North American canal era, travelers, as well as trade, continued to utilize these two well-trodden paths. During the canal's heyday, these two routes were followed on the popular Northern Tour that crisscrossed the northern border between New York State and neighboring Canada, and would markedly contribute to the making of a trans-border tourist industry in the Niagara–Great Lakes region.[14]

Eighteenth- and nineteenth-century travel accounts provide insight into the challenges of crossing either of these two routes in transit to Niagara, and the need to improve transportation in the Niagara–Great Lakes Basin. As early as 1750, Swedish explorer and botanist Peter Kalm wrote that he finally came to Fort Niagara "after a fatiguing travel, first on horseback through the country of the Six Nations, to Oswego, and from thence in a bateau upon Lake Ontario" to Niagara Falls.[15] Traveler Ralph Izard disembarked from a sloop in Albany in 1765, then took a wagon to Herkimer before boarding a bateau "rowed by poor soldiers" up the Mohawk River. Next he rode by horseback to Wood Creek, boarded a bateau on Oneida Lake, and rowed down the Oswego River to the Onondaga Falls where their boats "were all drawn out of the water, and rolled twenty yards, upon logs, made for that purpose below the Falls." Having next arrived at Lake Ontario at the mouth of the Oswego River, Izard transferred to a lake vessel to Fort Niagara at the western end of the lake where he finally mounted a horse and rode to Niagara Falls.[16] The Oswego–Lake Ontario channel that Izard followed was so well known and crossed that it would become a major contender for the route of the Erie Canal during the next century. Voyager Anne Powell's 1798 travel account provides insight into the alternate Montreal to Niagara route, in addition to the tourists' early interest in canals and other internal improvements as popular attractions *in their own right*. In total, the journey took about two weeks and Powell remained surprisingly good natured and upbeat considering the long voyage. Arriving in Montreal by stage coach, Powell and her party boarded a boat on the St. Lawrence that "kept along the shore" and was "obliged to go on very slowly." The next day "we reached a part of the river where our boats were unloaded and taken through a lock, the rapids being too strong to pass. These were the first rapids of any consequence I had seen."[17] Powell was referring to the historic lock at Coteau du Lac, built in 1779–1780 that was the first built lock canal in North America and a popular late eighteenth-century tourist attraction.[18] Powell also commented on the lock-keeper who happily offered an informal guided tour of the historic locks. The accommodations and mannerisms of the stage, boat, and river men who guided her party on their inland journey were also of interest. The boat she sailed on was small but neatly "fitted up with an awning to protect us from the weather . . . it was filled, eighteen persons in all, so you may suppose we had not much room." The crew ate and slept separately: "our sailors, they went on shore and boiled their pots, and smoked their pipes."[19] Powell mused that they "were encamped near enough for us to hear their singing and laughing" indicating that the quality of travel was often in the hands of the crew. Though not all accounts were as complimentary as Powell's, travelers were entertained by the river and boatmen who had to employ every ounce of energy and muscle to overcome hazards like sailing against rapids or hacking through bushes and overhanging tree branches from the river's edge.[20] On the tenth day Powell's party reached the provincial port of Kingston at the mouth of the St. Lawrence that would come to be an important commercial hub in the interconnected canal system of New York State and Upper Canada. Having rested at Kingston, Powell's party boarded a vessel on Lake Ontario in transit to Niagara Falls. During the next century, hundreds of northern tourists like Powell would come to Niagara not only to view the majestic Falls, but to witness some of the boldest innovations in canal technology and development.

The weak state of transportation and communications facilities inland during the 1790s encouraged Canadian and American politicians, merchants, land speculators, and business leaders to consciously promote internal improvements that would make the region more accessible while also promoting cross-border commerce, tourism, settlement, and expansion generally. A system of interconnected roads, canals, and other internal improvements would secure stability and prosperity in the Niagara region. Late eighteenth-century travel accounts and sundry sources indicate that improved roads were important conduits in the movement of people, goods, technology, and information across the Niagara borderland. The major road linking the Canadian and American settlements during the 1790s was the *Grand Thoroughfare* or State Road that provided access from Albany to Niagara, which then connected with a road to Buffalo or another "by the Tanawondoe (sic) Village to Queenston," Upper Canada. By the following decade, the State Road was used enough "by people on business, or by those whom curiosity had led to visit the Falls of Niagara, that a station was fixed . . . to shelter travelers," indicating an early cross-border tourist trade at Niagara.[21] Meanwhile, Holland Land Company agent of western New York, Joseph Ellicott, who arrived in the region in the 1790s, encouraged the building of roads leading directly to Buffalo and Upper Canada in order to promote the commercial and recreational potential of the region while also providing conveniences to settlers on the western purchase who looked to Canada for social and economic contacts. Of early interest to Joseph Ellicott and his brother Benjamin was the construction of a road from Buffalo that connected Benjamin's mill and tavern to the Niagara River and the nearby settlements of Upper Canada. Joseph Ellicott believed that both the mill and tavern would serve as important commercial and recreational linkages across the porous border.[22]

The north-south orientation of roads in the Niagara region illuminates the growing importance of an American-provincial transportation network that carried information, commerce, and people back and forth across the border.[23] As early as 1793, the American postmaster general proposed a postal route to Niagara to better ensure the carrying of mail to the northern border and "a connection with our neighbors in Canada."[24] The following year the *Upper Canada Gazette* announced "with pleasure" that "large sums of money are raised and raising for improving and extending the roads" from the American side to this place.[25] By 1812, the *Grand Thoroughfare* was providing a great network of roads between Canada and the United States, further illuminating the role of transportation improvements in facilitating an integrated borderland economy in this porous region:

> [F]rom a reference to the map of the State, it will be observed that this road is part of the great leading state road from the City of Albany to Buffalo Creek, at the east end of Lake Erie, where it communicates with roads to Upper Canada, Presque Isle and other places on the Lake—it is also the only route which is traveled by an increasing concourse of people, who are attracted by that great object of natural curiosity, the Falls of Niagara.[26]

The most vital road was the portage road around Niagara Falls that linked Lakes Ontario and Erie and points further west. For several years after the American Revolution, the British controlled the old portage on the east bank (American side) of the Niagara River, a road that ran eight miles between Lewiston, below the Falls, and Fort Schlosser above. But in

1791, the British transferred the portage to the western or Canadian side of the Niagara River, resulting in a slightly longer route. The new road was nine miles long, running from Queenston, up over the Falls to Chippewa on the Niagara River and the Lake Erie outlet. After the British moved the portage to the Canadian side, it became the preferred route of travel and commerce until the old portage on the east side was revitalized during the next century by Augustus and Peter Porter, two of America's leading public citizens whose investments in transportation, trade, and tourism helped forge a borderland economy in the Niagara–Great Lakes Basin.

A jaunt along the Niagara portage road was highly anticipated by tourists and travelers alike. It was easiest, and most pleasant, to reach the portage road by sailing first up the Niagara River before disembarking at either of the lower landings at Queenston, on the Canadian side, or Lewiston on the American side. Next, travelers faced a challenging climb up the steep slope of the Niagara escarpment or mountain before the portage road flattened out to the Falls. Though most people ascended by foot, oxen and horses pulling wagons were used to haul goods and luggage up the mountainside. At times, as many as six oxen were needed to haul a heavy load up the escarpment. An important improvement that alleviated the work-load of man and beast, while also providing a source of entertainment for tourists, was a device called "The Cradles," invented at the Lower Landing near Lewiston during the 1760s to bring supplies up the steep mountainside.[27] Particularly useful for hauling up heavy provisions and boats, this ingenious device was thought to have had its origins in the developing canal systems of England and Europe.[28] Fascinated by these improvements at Niagara, traveler Captain John Enys in 1787 described the cradles as a kind of "Cart on Small Rollers, which ran in Grooves. At top is a windlass by which they hoist the Cart up, with a great weight of goods on it, with a very small number of men."[29] Anne Powell watched with equal fascination as her luggage was "drawn up a steep hill in a cradle, a machine I never saw before."[30] One of the earliest internal improvements at Niagara to facilitate commercial and pedestrian traffic, the cradle or inclined railroad is believed to be the first of its kind in North America.[31] It was also a success as a tourist attraction as many traveler accounts from this period frequently referenced the novel device.[32] Several years later, future Erie Canal founder De Witt Clinton visited the cradles and was intrigued by the simple device that conveyed goods up and down the mountainside.[33]

After ascending the steep mountain, people crossed the nine-mile portage on foot, horseback, carriage, or sled in winter. Coming to the Falls in 1791, Viscount Chateaubriand arranged to procure his own "guide and some horses" at Albany before setting out for his Niagara adventure, indicating a scarcity of transportation facilities as one traveled further inland.[34] Local frustrations associated with travel sometimes occurred as suggested by Anne Powell whose party, after leaving the American side of the Falls, was forced to travel to "Ft. Schlosser by any means we could." Thankfully, "two Calashes were obtained—my brother drove in one . . . me in the other." Powell was describing one of the most popular eighteenth-century vehicles at Niagara—the calash—a two wheeled horse-drawn carriage with a folding top. According to Powell, the horses pulling the carriages were not in the best of condition and appeared "tired."[35] Meanwhile, an announcement that a child was killed "by the wheel of a wagon over him" illuminated the risks involved in traveling the portage road by horse and wagon.[36] Indeed, part of the popularity of canal travel during the next century was the belief, at least at first, that this new mode of conveyance would be less prone to dreadful accidents that frequently occurred on wagons and carriages, while also eliminating the excessive ware and abuse of horses that came with long-distance land transportation.[37] As both pedestrian and commercial traffic increased at Niagara, more reliable and sound-worthy transportation facilities were mutually undertaken by Canadian and American business leaders and entrepreneurs in an effort to strengthen the borderland economy and stimulate commerce, tourism, and development along their shared border.

Niagara's business leaders and developers from both sides of the border learned early on that visitors to the Falls were interested in roads and other internal improvements not simply for convenience sake but as *attractions in themselves*. For many a traveler, the thought of experiencing the latest transportation and communication innovations at Niagara were as exciting as the prospect of witnessing the majestic Falls. The portage road was a favorite among Niagara travelers who eagerly anticipated a ride along this scenic tongue of land. Military officers and traders regularly crossed the portage road while wagons laden with food, furs, and military supplies were a frequent and interesting sighting. After disembarking at Queenstown, John Maude observed with fascination the commotion along the portage road "which employs numerous teams, chiefly oxen; each cart being drawn by two yoke of oxen, or two horses. I passed great numbers on the road taking up bales and boxes, and bringing down packs of peltries." Maude felt like he had "suddenly been put down in Fairyland, I could not have been more alive to expectation."[38] By the early 1790s, the portage road was already regarded as one of the most beautiful drives in North America, and for the expectant tourist, a ride along the scenic portage could elicit as much emotion and response as that of the majestic Cataract. "Nothing," wrote one visitor, "can be more romantic than a road that leads from it [Niagara] to a place called the Landing about nine miles distant . . . in the summer evenings, it is the usual resort of those who seek air and exercise, and aided by the mild radiance of a setting sun, takes at every open, landscape worthy of the pencil of Claude."[39] A visitor traveled the same path in 1809, finding "the road excellent" and "the ride along the Niagara beautiful."[40] A year later, Canal Commissioner De Witt Clinton, while surveying the possibility of a canal around Niagara Falls, referred to the road as "pleasant and well-cultivated."[41] More than two hundred years later, thousands of visitors still come annually from the farthest stretches of the globe just to drive down this scenic stretch of road.

Borderland merchants and businessmen continued to make improvements to and around the Falls in an effort to bring more visitors to the region and stimulate tourism and trade. Though primitive by today's standards, even a simple improvement like a ladder down the side of the gorge offered tourists the best view of the Falls, or the opportunity to board a ferry or private boat to visit the American side. Without ladders, getting to the bottom of the gorge was a dangerous but thrilling proposition. Along with several other gentlemen in 1787, Capt. Enys, made the daring climb *without shoes* "as in many places it is so very slippery it would have been more dangerous to attempt with them on." Apparently, not all the men in his party made it down and those that did "candidly allowed . . . they would no means attempt it again until ropes or something more secure were placed in the most dangerous parts, as in some of the steepest parts they were obliged to let themselves down by means of a twisted stick."[42] Other tourists hired guides to take them down the shaky ladders showing that such primitive improvements were already attracting tourists in the eighteenth century.[43] Concerned about his young wife's safety, Simcoe had a ladder installed in 1795 so his wife could venture down the side of the gorge without fear of accident or injury.[44] In addition to providing a tourist novelty, innovations like "Mrs. Simcoe's Ladder" foreshadowed other improvements like stairways, roads, ferries, bridges, canals, and later railroads that helped forge an American-provincial borderlands network in the Niagara–Great Lakes Basin.

Next to Simcoe's ladder, the ferry was one of the earliest and most utilized forms of transportation at Niagara, bringing tourists, local residents, merchandise, and information back and forth across the Niagara River while providing an essential link between the Canadian and American sides. Until the nineteenth-century canal age, when ferry operators were forced to improve the quality of their service because of the sting of competition, ferries were basically row boats with no shelter or protection for passengers. By the mid-1790s, there were three major ferry crossings at Niagara—one between Fort Erie and Black Rock, one between Chippewa and Fort Schlosser, and one below the Falls at the landings.[45] Tolls were collected on the ferries but were generally regarded as an impediment to cross-border trade so for six days during the October *Annual Fair* in Upper Canada, American passengers, cattle, horses, and produce were admitted into the province free as both a gesture of good neighborhood and to stimulate business across the border.[46] A similar tradition holds today that allows passengers to cross the Niagara borderland free of tolls on Christmas day.

Tourism was stimulated at Niagara by the ferry service that catered to both sides. For many a tourist, the anticipation of crossing the tumultuous Niagara River in a ferry elicited the same feeling of terror and excitement as that induced by the great Falls. A gentlemen crossing as early as 1768 was greatly relieved that the ferry was "conveyed by Five sturdy men" who "conducted me thither in a bateau, and back again with safety, keeping dexterously between the two streams that rush on each side, to the Falls."[47] Simcoe's wife advised that great caution be taken in ferry crossings above the Falls, because "if they did not make exactly the mouth of the Chippaway," the force of the water below "would inevitably carry them down the Falls."[48] An account of several people perishing over the falls because of the "inexperience of the helmsman" confirmed Simcoe's worst fears but ironically also contributed to the excitement and terror of the Falls that drew visitors to the region.[49] Christian Shultz for example, visited the Falls in 1807 and was fascinated by how the ferrymen managed the strong currents. Apparently the key was to find "a small space in the river, over which a boat might cross with the greatest safety."[50] While visiting Upper Canada, Canal Commissioner De Witt Clinton was similarly struck by the art of ferrying the Niagara River: "We crossed the ferry at Queenston, which accords a curious phenomenon. An eddy runs up on each side, and facilitates a passage against a very impetuous current in the center of the river."[51] Smooth ferry crossings clearly depended on the expertise and fitness of the boatmen who skillfully weaved in and out of the rivers eddies before making a dash to the other side. Such expertise was especially relevant during winter crossings when the river was laced with ice. Canada's father of the Welland Canal, William Hamilton Merritt, lost his eldest sister and niece in a dreadful ferrying accident between Lewiston and Queenston after the boat was hit by a large piece of floating ice and capsized.[52] Merritt's interest in later years in building a suspension bridge across the Niagara River in place of the topsy-turvy ferry crossings must have partially been influenced by this dreadful accident.

The potential of the ferry to bring economic and recreational prosperity to both sides of the border was early recognized by local hotel operators and entrepreneurs. In 1797, a Canadian innkeeper announced that he had opened a "House" near the Fort Erie ferry promising "gentlemen traveling to, or from, the United States, or going over the lake," fine entertainment and accommodation.[53] When annual horse races were introduced at Niagara, Upper Canada, in 1797 so as to encourage "a greater circulation of specie and an intercourse of commerce, friendship and sociability between the people of this province and the neighboring . . . United States," it was the ferry that provided passage for both man and beast.[54] Another visitor to the region observed that the ferry was "the main channel of communication" between Upper Canada and the United States, in addition to promoting "a great intercourse between the two sides."[55] For Upper Canadians, the ferry offered another purpose—it delivered the news more rapidly than that which came down the Montreal–St. Lawrence outlet, so that Upper Canadians came to rely on American newspapers for their daily information.[56] The importance of the ferry in conveying information from the United States to Canada was revealed in Niagara's *Upper Canada Gazette*, the leading newspaper in the province in the 1790s, which contained more stories on the United States—many of which were plastered on the front pages of the newspaper while British news was often consigned to the back pages.[57]

The ferry also served more nefarious purposes as indicated by the extensive smuggling activities in the Niagara–Great Lakes borderland vicinity.[58] From the earliest times, tourists, local residents, and officials commented on the illicit trade, suggesting the overall importance of smuggling to the economic health of the borderland economy. Contraband across the lakes was a persistent problem for British and American officials but there was little that could be done to prevent the illicit trade. A story told in 1794 attested to the difficulty of trying to enforce regulations across the Niagara border. Apparently, three men carrying contraband in a boat were called ashore by Canadian officials but they refused to come in and shot at one of the guards, indicating the extent to which smugglers would go to ensure the illegal trade.[59] Rochefoucauld-Liancourt commented on the persistent problem at Niagara stating, "This contra-band trade will be a constant object of dispute between the two states" but not for the local residents who profited handsomely from the exchange.[60] An early Upper Canadian newspaper acknowledged that enforcing regulations between the two peoples was a[n] onerous and precarious mission.[61] A few years later, while touring Niagara Falls, a British traveler observed "smuggling was carried on from this country to Canada to an immense amount," especially since the embargo was laid.[62] Visitor John Melish professed that "smuggling had become so habitual, that it will probably give much trouble to the general government to prevent it."[63] The extremes to which local people would go to ensure the illegal cross-border trade was illustrated in an 1808 account in which an Upper Canadian merchant employed a local agent in the "smuggling business" to tie up a Lewiston deputy sheriff so that his large order of tea and other goods from New York City could be brought across the Niagara River.[64] Though smugglers left little in the way of quantifiable evidence, such accounts reveal the constant flow of tea, coffee, tobacco, and other goods across the border that helped sustain the borderland economy.[65]

Indeed, smuggling was so prevalent in this region that a hotel was established at Black Rock, opposite the Fort Erie ferry that served as a cover to the illicit trade (fig. 2.1). A fine tourist establishment by day, at night the hotel acted as a cover for wagonloads of flour and other merchandise to be conveyed across the river into Canada.[66] The story of a fellow traveler

Distant view of Black Rock and vicinity.

Figure 2.1. **Black Rock, New York.** As early as the eighteenth century, Black Rock served as a major crossing point to Canada. During the War of 1812, Black Rock became a notorious smuggling center that sent much produce to the Upper Canadian and Montreal markets. During the canal age, Black Rock also became a well-known crossing point to Canada on the infamous Underground Railroad. John W. Barber and Henry Howe. *Historical Collections of the State of New York.*

who, while sitting aboard a schooner on the Niagara River heard customary toasts made "first to the King, and next to President Jefferson" (the latter nod was presumably in recognition of the profitable cross-border trade since Jefferson's embargo was enacted), provides evidence of a sustained cross-border trade even if illegal.[67] Again, visiting the area before the war, John Melish believed that the widespread practice of smuggling would not only give "much trouble to the general government to prevent it," but "may in fact be productive of confusion and bloodshed." De Witt Clinton also recalled that during the war smuggling was facilitated by a ferry below the Falls.[68] When peace finally returned to the frontier in 1815, local entrepreneurs found new ways of capitalizing off the ferry which, in conjunction with canals and other internal improvements, drew more and more people to the region. As part of the Northern Tour at Niagara during the next century, horse-driven ferries would take visitors on tours of the newly erected canal sites and related attractions that helped transform Niagara into the tourist mecca of the world.[69]

Improvements to roads, walkways, and carriage and ferrying services overlapped with the beginnings of a transnational economy in the Niagara–Great Lakes borderland region. A spirit of cross-border community was early forged by businessmen and developers like Benjamin Barton of Porter, Barton and Company that took over the portaging of goods around the American side of Niagara Falls after 1806. Barton first came to the American side of the river in the late 1780s and organized a land company with several British provincials constituting one of the first borderland ventures in the region.[70] Migration, particularly the large influx of late loyalists into Upper Canada also promoted a borderland economy. Migrants traveling west and into the upper province had cash and western New Yorkers helped meet their needs by offering supplies, shelter, taverns, fodder for animals, and other services. "The amount of money received from these sources is difficult to estimate, but certainly it was considerable" writes historian Neil McNall in his study of New York. As stated earlier, McNall noted that migrants, dependent settlers, and newcomers to Upper Canada may well have "consumed more goods than were marketed via Albany, the Susquehanna, and the St. Lawrence before the opening of the Erie Canal."[71] Reflecting on the late loyalists' impact on the upper province, traveler Isaac Weld wrote in 1796: "So sudden and so great has the influx of people, into the town of Niagara and its vicinity, been, that town lots, horses, provisions, and every necessity of life have risen, within the last three years, nearly fifty percent in value."[72] Across the Niagara River, Holland Land Company agent of Buffalo,

Joseph Ellicott, pushed for an immediate liberalization of land company policies so that more people might settle in the American purchase. The company's easier land terms encouraged some British subjects, including Americans, to return to Buffalo, indicating the ease in which people moved from one country to the other and the permeability of the border in this region.[73] Traveler La Rochefoucauld Liancourt witnessed the back and forth movement of people, stating that large companies of American emigrants to Upper Canada were considerable, but "emigration from Niagara to the United States is also considerable."[74] As observed by historian Reginald Stuart, these experiences created a "genuine borderland where local loyalties based on human relations, survival, and self-interest often transcended allegiance to a distant national government."[75]

Other stories attest to the growing importance of an integrated borderland economy at Niagara. During the late eighteenth and early nineteenth centuries, a sustained flow of money, goods, animals, and people moved back and forth across the border.[76] In the immediate years after the American Revolution for example, a hefty sum of money was made by American merchants crossing cattle from Lewiston, New York to sell in the Upper Canadian market. A story is told that in the late 1780s so many migrants, in addition to "drovers with considerable sums of money" crossed the border at Lewiston that it attracted the attention of robbers.[77] The cattle were sold principally to Upper Canadian settlers and to the garrisons at Queenston and Niagara. American goods, especially cattle, horses, and other provisions that were too expensive or difficult to transport from the east, were welcomed in Upper Canada.[78] In 1792, an American resident living between the Genesee River and Fort Niagara on Lake Ontario started a prosperous business buying and marketing fish in Upper Canada. Having obtained a boat, he crossed Lake Ontario to the "river Credit"[79] where fresh salmon was purchased, then carried it back to the American settlement. Here he exchanged the salmon "for butter and cheese which he would market in Canada, making large profits."[80] Salmon was a profitable industry for both Native residents and whites, it being recorded by more than one individual that salmon "swarmed the rivers" and lakes so thickly that they could be caught with a shovel or even by hand.[81] Other goods like whiskey and rum regularly crossed from New York to the Canadian side. In the 1790s, a merchant having just arrived in the Genesee country built "A convenient store house" (near the mouth of the Genesee river) which during "the last two summers, very considerable quantities of provisions and distilled liquor were sent from this place to Canada."[82]

Niagara's settlers found "at an early period, the most advantageous markets for their surplus produce. To Canada, beef, salt, pork, flour, and whiskey, are already sent to a great amount."[83] In 1797, an Upper Canadian merchant at Niagara announced that he had a fresh supply of goods "From New York and Albany" including cotton, tobacco, handkerchiefs, and clothing.[84] Another Canadian shopkeeper advertised that he had for sale a variety of "American cheeses" that could be complimented with fine brandy and "Jamaican spirits."[85] A visitor to the region in 1799 was impressed by the level of economic activity across the border, saying "the early settlers . . . find the most advantageous market for their produce in Canada, where they send their beef, pork, flour and whiskey."[86] During the same time, La Rochefoucauld Liancourt observed that "[m]any commodities, together with numerous droves of cattle, are exported hence (from western New York) annually into Upper Canada."[87] Settlers on the American side of the river were also dependent on Upper Canada for markets, wares, agricultural products, grist mills, and other sundry products.[88] This was illustrated by the day-to-day happenings of men like T. S. Hopkins and Otis Ingalls who were the first to raise wheat on the Holland Purchase near Clarence, New York, and when it was ready for grinding transported it to Chippewa, Upper Canada: "They went with three yoke of cattle, by way of Black Rock," where they caught the ferry to the other side.[89] As observed by one scholar, "Americans and Canadians crossed the boundary to have their grinding done in the nearest mills and all of them prepared potash and timber" for sale in local and more distant markets.[90] The degree of back and forth commerce was significant with the border remaining "permeable to both populations and trade."[91] Though customs houses were established on both sides of the Niagara River by 1802, Jane Errington notes that even "the levying of customs duties after 1800 and the imposition of an American embargo in 1807" did little to hamper the north-south trade that was already "one of the most important sources of prosperity to the colony."[92]

Local farmers, merchants, businessmen, and politicians profited from the brisk cross-lake trade.[93] Even famed New Yorker Alexander Hamilton, who served as America's first Secretary of the Treasury, recognized and promoted cooperation between the neighboring countries. In the early 1790s, Hamilton predicted the advantages of an open and reciprocal trade with Canadians, a people who were growing in wealth and numbers: "in Upper Canada there is no reason to believe that the future progress will be slow. In time to come the Trade may grow into real magnitude." Hamilton argued that restrictive trade laws only encouraged contraband: "Entirely to prevent trade between bordering territories is a very arduous perhaps an impractical task." Hamilton concluded that the border could never be closed to either smugglers or honest traders whose livelihoods depended on the northern exchange, while acknowledging that trade with Canada was "essential" to the interests of New York State.[94]

Like Hamilton, travelers to the Niagara region commented on the economic and social bonds between the neighboring countries. While visiting in 1793, one traveler was struck by "the greatest harmony" that exists "between the American and British subjects in that quarter."[95] A 1795 poem about Niagara indicated that no amount of national legislation or coercion would impinge upon the reciprocal trade:

> Here, where no laws mankind refrain,
> But only what themselves ordain,
> Where unrestricted commerce wide,
> Pours in her wealth with every tide. . . .[96]

Indeed, by the time British and American officials signed Jay's Treaty in 1794, free trade was already a reality at Niagara. The treaty recognized "neutral reciprocity of trade" and granted to each other's nationals "the right to pass freely over the border and to make free use of the lakes, rivers and carrying places on both sides of the boundary for commercial purposes."[97] Historian John Bartlett Brebner notes that along the New York–Canada borderland the reciprocal terms of the treaty were "already the normal procedure in the form of practically free trade between American and Canadian frontiersmen."[98] Upper Canadian politician Peter Russell celebrated "The many reciprocal advantages which the inhabitants of Upper Canada and their neighbors in the United States may derive from this treaty" that will "conciliate and preserve "the most perfect good understanding and harmony between the two countries."[99] A passing American traveler in 1799 agreed, noting with "Goods on importation being liable to no duty," this country will have "a vast advantage . . . indeed nature points out this place as the emporium of trade for the people inhabiting both sides." The potential for growth in this region seemed limitless to borderlanders engaged in the trade: "The Straits of Niagara, from its peculiar situation, being the channel through which all the produce of the vast country above must pass, it is looked forward to as a place of the first consequence."[100]

In light of Jay's Treaty and the growing commercial and pedestrian traffic along the northern borderland, the need for a canal around Niagara Falls seemed evident. Indeed, as early as the seventeenth century, travelers commented on the usefulness of boat navigation between Lakes Erie and Ontario that would eliminate the portage that carried goods and people above the Falls and points farther west. Famed traveler, Father Louis Hennepin, who is best remembered for bestowing Europeans with their first visual and written accounts of Niagara Falls, reflected on the need for such a canal. Arriving on a brig of 10 tons at the mouth of the Niagara River in 1678, Hennepin was part of LaSalle's expedition, another famous explorer who built the first European vessel, the *Griffon* on Lake Erie.[101] In his *New Discovery of a Vast Country in America* (1699), Hennepin writes of the "four great lakes" that descend upon "the Great Falls, which is the continuation of the great river of St. Lawrence." "If not for this vast Cataract," he observes, "which interrupts navigation, they might sail with Barques or greater vessels above four hundred and fifty leagues further, cross the lake of Hurons, and up the further end of the Lake Illinois."[102] Anticipating the canal era by some one hundred and fifty years, Hennepin projected the need for a ship canal around Niagara Falls that would open the interior to the Atlantic world of commerce and trade. Several years later, French commandant Antoine de la Mothe Cadillac commented on the utility of such a project. Interestingly, Cadillac's wife was the first known white woman to make the arduous crossing in 1701 and, like her husband, must have appreciated the idea of a more efficient means of transit. Around the same time, her husband enthusiastically speculated that Niagara's navigational impediments could be overcome, and trade increased, by making a junction or canal "between Lakes Erie and Ontario."[103]

As traffic increased at Niagara, it became necessary to introduce more reliable and sound-worthy transportation to and around the Falls. The commercial and recreational success of the portage spurred visions of a grandiose canal skirting Niagara Falls between Lakes Ontario and Erie that would improve transportation and communication facilities, spur market development, tourism and expansion, and promote greater reciprocity of commerce and interest between the neighboring countries. As early as 1772, the authors of *The Rising Glory* saw Niagara's future in such a canal:

> Hoarse Niagara's stream now rolling on
> Thro' woods and rocks and broken mountains torn,
> In days remote far from their ancient beds
> By some great monarch taught a better course,
> Or cleared of cataracts shall flow beneath
> Unincumber'd boats and merchandise and men.[104]

A few years later, a traveler similarly commented on the necessity of the Niagara canal observing that while the Niagara rapids and falls "are the greatest natural curiosities in the world, nature seems cruelly to have intercepted the navigation of this vast Mediterranean of America, and the commerce of this part of the world will belong, perhaps forever, retarded by this immense cataract of water."[105] It was precisely because of the growing commerce and travel at Niagara that American and Canadian transportation leaders set themselves to the task of overcoming Niagara by boat or sloop navigation.

Few people came to the region without speculating on the usefulness of a canal around Niagara Falls. Civil Engineer William Tatham was one such visionary. Born in England during the 1750s, he moved to America and served at Yorktown during the American Revolution.[106] A proficient writer, he penned essays on canals and inland improvements, and during the 1790s returned to England where canal construction was in its heyday. While in London, Tatham connected with Robert Fulton, who is credited with building the steamboat *Clermont*, and his associate Joshua Gilpin, all of whom were caught up in the canal craze.[107] There Tatham began writing on canals and locks, but it was the idea of an inclined plane that interested him the most. In fact, one of his more grandiose designs in the 1790s was for an inclined plane that would carry large vessels over Niagara Falls between Lakes Ontario and Erie.[108] Tatham's Niagara project would transfer "any vessel whatsoever, or however great may be her burthen, for which there is a depth of water to come up the rivers Hudson and St. Lawrence, to the Falls: or sufficient surroundings in the waters above to answer for the largest possible sized barks, which can also navigate the upper lakes."[109] His project promised to make Niagara Falls

> the commercial center of all the surrounding countries of the lakes, and the grand mart for the raw and rough prepared materials of interior American produce, and the political center where the provincial interests of Great Britain and the United States should be made to harmonize, and meet each other in mutual offices of brotherly love, and reciprocal advantage.

Though his project was never realized, Tatham believed that a canal connecting Lakes Erie and Ontario promised enormous commercial and reciprocal benefits to both countries.

The last two decades of the eighteenth century saw important advances toward improvements inland, including the idea of a Niagara ship canal. Between

1784 and 1786, Christopher Colles, a native of Ireland who settled in New York before the Revolutionary War, in association with assemblyman Jeffrey Smith of Long Island, proposed a number of bills to the New York legislature for the removal of obstructions on the Mohawk River between Albany and Oswego, with the intention of extending these navigational improvements to Lake Erie by way of a canal around Niagara Falls.[110] Although little came of these initial efforts, Colles's enthusiasm for "connecting the amazing extent of the five great lakes, to which the proposed navigation will communicate," so resonated with canal minded individuals that in the 1790s improvements to navigation on the inland rivers and lakes west of Albany were undertaken.[111]

The year 1792 proved a critical milestone in the history of internal improvements inland. It was during that time that the private Western and Northern Inland Lock Navigation Companies were incorporated, the former to improve navigation between Albany and Lake Ontario, and the latter to open navigation between the Hudson and Lake Champlain. The Western Company undertook the canalization of the Mohawk-Oswego water system to promote commerce and travel in the Great Lakes region. As one scholar observes "this, the first man-made transportation route from the Atlantic to the Lakes, was international in character—for the Fort at the mouth of the Oswego River was still in the possession of the British . . . and the westward bound traveler from Fort Oswego skirted the southern shore of Lake Ontario to the Canadian side of the Niagara River, and their disembarking followed the road across the Niagara peninsula to Lake Erie."[112] Backed by private financiers, speculators, and wealthy property holders, the Western Company effected dramatic improvements to navigation between Albany and Lake Ontario. Standing on the southern shore of Lake Ontario, Englishman and company engineer William Weston, reflected on the limitless potential of this natural trade corridor:

> . . . it is almost superfluous to remark (what is so obvious to every person the least acquainted with the geography of the state) on the immense expanse of internal navigation that opens upon our view; the extent of these lakes . . . presents the mind a scene unequalled in any other part of the globe, offering to the enterprising and adventurous sources of trade rapidly advancing to an incalculable amount.[113]

Serious efforts toward realizing a canal at Niagara came in the wake of the Western Company's impressive advances to inland navigation. Taking what was once an obstructed and interrupted navigation along the Mohawk River between Schenectady and Oneida, the company, through the construction of canals and locks, opened a relatively continuous deep-water channel for large boat traffic to pass.[114] Work crews came from as far as Canada, Vermont, Pennsylvania, and Connecticut to labor on the project, allowing improvements to move forward on the Mohawk River, despite the inevitable delays and difficulties associated with canal digging in late eighteenth-century America, while also early indicating that the Canada–United States border did not act as a barrier to Canadians seeking employment opportunities on the American canal system.[115] However, the opening of canals and locks at Little Falls, the German Flats, Rome, and Wood Creek offered considerable evidence of the commercial and agricultural benefits of canals and other such improvements to the state.[116] "So great were its practical results," wrote a friend of internal improvements, that by 1796, "boats of 16 tons burthen were enabled to pass from Schenectady to Seneca Falls, a distance of 212 miles, and the price of transportation was reduced from 100 dollars to $33 a ton." As the "price of transportation was reduced" the value of produce, and of lands in the vicinity "was increased, and the benefits otherwise resulting from the system, was justly appreciated by many public spirited individuals."[117] Meanwhile, the cost of transportation from Albany to Niagara Falls was cut in half.[118]

In "a country where improvements to water navigation" had "never been prosecuted to any extent," the importance of the Western Company improvements were not lost on canal leaders or passing travelers. As explained by one Superintendent in 1793, the undertaking was an historic achievement "it being the first work of its kind in the Country."[119] Traveling through New York State between 1806 and 1809, John Melish deliberately stopped at Little Falls to observe where the canal had been cut more than a decade earlier. Impressed by the work, Melish noted that "it was originally built of wood, but that falling to decay, it was rebuilt of stone 8 years ago. There are 8 locks at this place. The toll has been lessened within these years on account of the wagons taking away the trade from the canal."[120] When completed the canal actually consisted of five locks, the number of which were sometimes confused by visitors perhaps because of the sheer enormity or novelty of the project. Indeed, a traveler to Niagara in 1809 took a journey to the old canal noting that at Little Falls there was "a canal with six locks" and a bridge across the Mohawk River.[121] Still, as observed by Melish and other passersby, travel time and the costs of transportation between Albany and Niagara Falls were continuously being reduced as a result of these improvements. A major benefit of the Western improvement along the Mohawk-Oneida trade corridor was its natural connection to Lake Ontario and potential trade ties to Canada. The commercial and recreational significance of this route to the larger interests of New York State would be brought to the public's attention during the next century when the New York State Canal Commission undertook surveys for a canal between the Hudson River and the Great Lakes.

Just six years after the founding of the Western Inland Lock Navigation Company, an act of the New York Legislature called for continuing the line of communication around Niagara Falls "for the express purpose of connecting the vast chain of Upper Lakes, with Ontario and the Hudson, via the Oswego, and improved navigation." Closely identified with the Niagara Canal, and the state's internal improvements in general, was Elkanah Watson whose position as director of the Western Company furthered his zeal and enthusiasm for the Niagara project. Having spent much of his earlier life touring Europe and the United States, Watson kept numerous travel records, consciously detailing every canal he came across on his journeys. Watson's passion for canals and other transportation systems brought him to New York in the 1790s where the canal mania had just gotten underway.

As a traveler, writer, and visionary, Watson understood the importance of internal improvements to the agricultural, commercial, and recreational wellbeing of a developing economy. In addition to his efforts on the Western Canal, Watson faithfully promoted the idea of a canal around Niagara Falls. Believing in the practicality of utilizing natural rivers and water courses, he wrote "the utmost stretch of our views, was to follow the track of Nature's canal, and to remove natural and artificial obstructions. . . ."[122] Watson's son confirmed that from the incorporation of the Western Company in 1792, the intention of the men with regards to the improvement was to connect the Lake Ontario route "with the waters of the upper lakes, by the construction of a canal around Niagara Falls."[123] With much vision and energy, Watson devoted himself to obtaining the passage of a law to open a canal and locks around the great Falls of Niagara. For Watson, one had only to look to the geographical position of the continent with its hundreds of miles of virtually uninterrupted boat navigation between the upper Lakes and the Atlantic to realize the vast potential of the Niagara project. Identifying himself as the *Anticipator* in the *Albany Register*, Watson prophesized:

> Let us suppose the canals in this state completed to Oswego (for bateau of 15 to 20 tons burden), a fine natural harbor on the south side of lake Ontario, where a town is soon to be laid out; here a sloop of 70 tons will take in her cargo and proceed through the lake Ontario to the Niagara canal—thence to Lake Huron to the falls of St. Mary, or thro' the streights (sic) of Michilimackinac to the south end of Michigan, in all 1300 miles in a direct line without interruption.[124]

Watson became animated when talking about "the wealth, population, civilization and power" that would emanate from this mammoth undertaking.

The private Niagara Canal Company was incorporated by an act of the New York Legislature on April 5, 1798. Besides Elkanah Watson, the chief movers behind the Niagara canal were James Watson, a friend of Elkanah's who was encouraged to invest in the project; Benjamin Prescott, a Massachusetts engineer who two years earlier had made an extensive survey of the Niagara Canal; Genesee promoter and land agent, Charles Williamson; and John Williams and Effingham Embree, all of whom possessed an ardent zeal for the advancement of internal improvements. Situated between the waters of Lake Ontario and Erie, the Niagara Canal promised to "facilitate and advance the internal commerce of this State; and provide the convenience and prosperity of the people thereof." The company directors were confident that the canal would promote commerce and agriculture and encourage settlement in the Great Lakes region. The company's vision was of a canal that would link the far distant regions in the East with the continental heartland in the West. Next to the commercial benefits of such a vast system of navigation, the recreational potential of the canal which "commenced above the rapids" and descended "in majestic confusion to the grand Cataract," would make it "the greatest display in nature. . . ."[125]

A dearth of evidence on the projected American Niagara Canal has resulted in little in the way of scholarly analysis. However, like much writing on the North American canal era, the Niagara project tends to be viewed, if considered at all,[126] as part of the broader imperial contest between the United States and Great Britain for dominance of the midwestern trade. The Niagara canal would serve as an extension of New York State's internal improvements scheme to divert trade away from Quebec and the St. Lawrence, and secure it for the New York City market.[127] However, for its earliest proponents, the Niagara Canal was not simply designed for advancing America's interests inland but rather as a means of overcoming Niagara's navigational impediments in the Great Lakes–Atlantic water system that hindered commercial and recreational growth and prosperity in this vital region. No one was more readily aware of the canal's value to both Canadian and American settlers than Niagara Canal Company promoter and Scottish immigrant, Charles Williamson, who first came to the region in the 1790s as an agent for the vast Pulteney estate, a large tract of land that encompassed much of western New York. A keen promoter of settlement and agricultural and commercial development in the western district, Williamson regarded closer social and economic ties with neighboring Canada as essential to the region's growth and prosperity. Indeed, Williamson takes credit for establishing one of the first mail and newspaper routes between Albany, Genesee, and Niagara, Upper Canada. Ever a friend of Anglo-American rapprochement, Williamson pushed for the opening of the international border between Canada and the United States three years before Jay's Treaty was formally ratified in 1795. In association with the Pulteney agents, Williamson also backed the development of Sodus Bay on Lake Ontario to encourage trade with Montreal, while eagerly promoting a water connection to Albany on the Hudson by way of Oswego.[128] In addition to the obvious commercial utility of canals and other internal improvements, Williamson saw in the Niagara project as many opportunities north and south as he did east and west.

With the backing of men like Williamson and Watson, several studies and surveys were undertaken to advance the Niagara Canal. Following these reports, it was determined that the canal would commence "[f]rom the most convenient place above the falls of Niagara at or near Steadman's landing to the most convenient place below said falls, and nearly opposite to Queenstown landing, and to construct in such canal . . . all such locks, dams and other works and devices as shall be necessary for the purposes of making a complete navigable water communication" between the two places. The canal would extend some seven miles in length, and would require forty locks with a descent of about 320 feet.[129] Benjamin Prescott had earlier calculated the cost of the canal at "six hundred and twenty three thousand dollars," if constructed for small vessels, and "if for sloops of seventy tons burthen," a million dollars.[130] As chief director of the company, Watson could not have been happier with the findings. In a bid to sell the canal to the public, he wrote that the cost was a small sum compared to the canals "in Europe and China, which cost infinitely larger sums." "I challenge the world," he continued "to exhibit a canal so important in its consequences, as is the one in question."[131]

From its early inception, the Niagara Canal Company faced hurdles and challenges typical of canal construction during the late eighteenth century.

A paucity of qualified engineers and investors with large reserves of capital was just some of the difficulties and problems that America's canal builders confronted. During the late 1790s, as reports of financial irresponsibility and fraud surfaced on other North American lines, confidence in canal construction waned. In an effort to ensure that the Niagara Canal would serve the common good, and to prohibit power from falling into the hands of a few large shareholders, it was determined that no one stockholder was "entitled to a greater number than thirty votes," with the number of votes being determined by the number of shares held in the company. However, a reference to defaulting and delinquent stockholders in the Niagara Canal Company charter indicates that as capital investments, private canal companies often failed. The prohibitive cost of the canal would be offset by allowing stockholders to collect tolls (though not to exceed $7 per ton on the tonnage of each boat or vessel), erect mills, build toll houses and other works, and to take from the land belonging to the state "such timber, stone and other materials as may be necessary to the construction of the said canal and locks."[132] A letter published by Elkanah Watson in the *Albany Register* suggested that the legislature had turned over several thousand acres of land to the company in the vicinity of the intended canal. If the canal and locks were not completed within ten years, the company would cease to exist and the lands would revert back to the state.

But similar to so many private undertakings during this time, the Niagara Canal did not get far beyond the planning stages. According to Robert Troup, who served next to George Washington during the American Revolution and later wrote a series of letters on the lake canal policy inland, the interest alone on the monies to construct such a work would have far exceeded the expected annual tolls on the canal.[133] The experience of the Western Company, though having made considerable strides in inland navigation at Rome, the German Flats, and Wood Creek, early proved that canals absorbed extensive capital with little monetary return.[134] Stories of corruption, miscalculations, and financial mismanagement on other American works must have weighed heavily on the minds of potential investors in the Niagara project. Nor did local residents always welcome the construction of canals and roads that could mean the appropriation of land by the state, the diversion of water from mills and streams, and loss and damage to property. While the Niagara charter specified that it was the company's responsibility to do "as little damage as possible" to the grounds and lands that they passed through, experience on other canals during the period revealed that local farmers and residents often incurred loss and injury to their property as a result of big companies incorporated to dig canals and roads in their district.[135] During the nineteenth-century digging of the Erie, Oswego, and Welland Canals, these complaints were magnified as farmers, property owners, and various settlers came to see that internal improvements did not always bring prosperity and progress in their wake.[136]

Reflecting on the Niagara undertaking, a passing visitor at Niagara Falls, John Maude, believed the canal a monumental task, especially given the difficulty of cutting through miles of limestone and fissures, making "it necessary to line the canal with tarred plank, or other materials impervious to the water."[137] Eighteenth-century canal building posed formidable challenges, and canalling Niagara would have been a herculean endeavor requiring huge capital outlay, and an experienced labor force. Indeed, though initial enthusiasm for such projects might be high, it was not until the Erie Canal's completion during the next century that qualified and experienced engineers would be available in North America to undertake such monumental works. The planning, surveying, and resurveying of the Niagara Canal alone suggested a lack of experience and confidence among America's canal directors and builders during the 1790s. Elkanah Watson admitted as much, stating years later that the reason canal companies so often failed during these early years was the fact that "we were all novices in this department."[138] In a last ditch effort to save the American canal, the Niagara Company had more surveys and estimates run but they were not carried into effect. Still, the Niagara Canal Company was one of the first enterprises that sought to open the far distant interior to agricultural and commercial development. Though the Niagara Canal Company failed to realize its ambitious project in 1798, the company representatives expressed a prophetic conviction "that in a few years this all important enterprise would be found both necessary and indispensable (sic)."

The influence of the American Niagara Canal in spurring transportation initiatives in neighboring Upper Canada became evident in 1799 when a bill presented to the Legislature called for an identical Niagara Canal only on the opposite side of the river. Introduced to the Assembly as "A bill to Improve and Amend the Communication Between Lakes Erie and Ontario, By Land and Water," the force behind the legislation was Robert Hamilton, who dominated the Canadian portage business after it was moved to the Canadian side in 1791.[139] Hamilton's forwarding and carrying business relied on good communication facilities and his enthusiasm for increasing transshipments across the Canadian portage encouraged him to push for a canal. In addition to supplying the British military and local economy,

Hamilton sold to American settlers across the river. As indicated earlier, Hamilton's preference was for American over British settlers who easily adapted to the North American environment. This view also promoted his business interests. The opportunity to increase trade with the United States excited Hamilton who wrote:

> The demand for Goods of all kinds from this Province must further increase with the progress of the American settlements that are forming along the south side of the St. Lawrence River and the Lakes. The natural, we may say only outlet for all produce of the settlements is by the River St. Lawrence. . . .

Hamilton understood the importance of the American settlement in populating his lands and selling his goods.[140] Indeed, a major market for Hamilton, outside of the British garrisons, was supplying the American garrisons with provisions, especially Fort Niagara that offered Canadian farmers a profitable market for their goods[141] and an early insight to the transnational economy in this region. In 1797, the demand of American troops for food like peas and flour practically exhausted the Canadian supply. Upper Canadian merchant Richard Cartwright noted at the same time that the American garrisons and settlements on Lake Erie and Ontario were completely dependent on Upper Canadian farmers for their bread. Some Upper Canadian farmers, who had contracted to deliver supplies for the British troops, sold instead to the American troops because they could demand a higher price. One scholar observes that Upper Canadians may have been taking advantage of the needs of their neighbors across the border but "there was nevertheless a community of interests between them. New Yorkers long had their grain ground in Canadian mills, flour and household articles were freely borrowed, and visits and gossip were exchanged." Meanwhile, "the Yankee peddler from Albany paddled his canoe along the shores of Upper Canada or New York with perfect indifference to the boundary line, for like the Yankee carpenter or mason, he received the same welcome at every clearance."[142] In 1796, Upper Canadian Road Commissioners, acting on behalf of Hamilton's portage company, pushed for further improvements to the portage road so as to "benefit the commercial intercourse of the country and add to the beauty of the place" further illuminating the vast economic and recreational potential of the Niagara borderland.[143]

Hamilton stressed the ties that united Canadians and Americans across the border rather than their differences.[144] Indeed, Hamilton regularly corresponded with Joseph Ellicott who during the early 1790s promoted the settlement of Buffalo on the American side of the Niagara border. Hamilton understood that the interest of his own locality on the Canadian side of the river would be "vastly benefitted by the settlement and improvement" of the neighboring United States. Hamilton was also closely affiliated with New York merchant John Porteous who operated a trading business at Little Falls on the Mohawk River. As previously discussed, it was at Little Falls that the Western Inland Lock Navigation Company completed its historic canal in 1795, which predated the Erie Canal by some 30 years and opened a major attraction on the Northern Tour. In addition to their shared interest in canals and commerce, Porteous regularly supplied Hamilton with products that included "mill stones from Schenectady" and "boulting [sic] cloth of the best quality." Other items shipped to Niagara, Upper Canada, by the New York firm included scythes, axes, tea, nails, candles, indigo, and French brandy. Hamilton understood that the interests of his own locality would be vastly benefitted by encouraging a north-south orientation of trade and transportation in the Niagara–Great Lakes region.[145]

Hamilton's pro-Americanism "expressed the general sentiment of the Niagara peninsula." With the proximity to the border, Upper Canadian contacts with the United States were extensive. In addition to buying oxen and horses from the American side that made possible the movement of merchandise on the portage, Upper Canadians purchased American-made wagons and boats because of their known durability and craftsmanship. From the "Pennsylvania wagon" that conveyed people and goods across the portage,[146] to the "Schenectady boat" that originated from Schenectady, New York, and the "Kentucky boat," a large covered riverboat popular in the western United States and early used in Upper Canada, Upper Canadian merchants and commercial operators relied on the American models.[147] Advertised in the *Canadian Constellation* as "three" or "five" handed for the number of men needed to pilot it, the Schenectady's flat-bottom and shallow draft was well-suited for Niagara's currents and rapids.[148] In 1799, Hamilton employed an American boat-builder to construct a vessel modeled after those used "on the Susquehanna and other rapid rivers." Compared to the boats already employed in Hamilton's service, the American model required fewer hands to man, and could carry 100 or more barrels of cargo.[149] Upper Canadians were early admirers of the American model. American influence on Canada's transportation developments continued well into the next century when skilled builders, engineers, laborers, and contractors were eagerly recruited from south of the border to work on the Welland Canal and St. Lawrence River system.

Like his American counterparts across the river, Hamilton thought that he could achieve the herculean task of building a canal above the Canadian Falls that would promote growth and settlement along the borderland. Hamilton was one of the first merchants in the region to support open and friendly trade with the United States, and there is no doubt that his plan for a canal included carrying American and Canadian produce.[150] In order to accommodate the growing North American trade, Hamilton called for improving the road between Queenston and Niagara above the Falls in addition to a canal "sufficient in width and depth to admit and allow any of the largest of the boats now used on that communication to pass" the rapids at Fort Erie.[151] The canal and locks would be privately undertaken by Hamilton and a handful of local merchants, and money raised through tolls would pay for operating expenses over a period of twenty-one years. Boats with goods and merchandise, "not having been weighed before passing the canal," would pay seven shillings and six pence "for each and every boat," but otherwise "the sum of two pence lawful money of this province" shall be paid on each and every hundred weight of goods, wares and merchandise." The total cost to build the canal was estimated at £50,000, and financial support was requested from the Upper Canadian government. The stipulation that the canal would operate "at all hours of the day and night, except Sundays, and except on Good Friday and Christmas day, when this canal shall be shut," early spoke to the belief that canals must be held to religious and moral accountability.[152] Though Hamilton's vision was never realized, the idea of a canal around Niagara Falls that would accommodate the British provinces and the United States was compatible with his pro-Americanism and emphasis on promoting a North American community in the Niagara–Great Lakes Basin.

Isaac Weld, after watching carts get hauled across the portage thought "it would be practicable to cut a canal" around the Falls like the one Hamilton envisioned. However, he also understood that such mammoth undertakings were cost-prohibitive, a problem confronted by late eighteenth-century canal promoters on both sides of the border. Weld predicted that a canal around Niagara Falls "will in all probability be undertaken one day or other," but that it would be on the American side and financed by New York State because they "are far better enabled to advance the large sums of money that would be requisite for cutting a canal . . . than the province of Upper Canada either is at present, or appears likely to be."[153] Weld was partially right in his prediction. Less than three decades later the Welland Canal—that skirted the Canadian side of the Falls between Lakes Ontario and Erie—was built thanks largely to New York State finance and resources.[154]

Though both countries earliest transportation visionaries and leaders understood that internal improvements were critical to the region's prosperity and future development, the boldest vision—the idea of a canal around Niagara Falls that would connect the Upper Great Lakes with Lake Ontario and the tributary waters below—would have to wait until the next century when North America's canal leaders sought to overcome Niagara's key navigational limitations in the Great Lakes–Atlantic water system. Like numerous private canal undertakings in the late eighteenth century, a dearth of skilled engineers and builders and insufficient funds retarded the Niagara project. Still, the vision of overcoming Niagara by means of a navigable water system sustained the interest of North America's most prolific canal leaders who during the next century went to work on behalf of the larger transportation revolution then sweeping over the continent. As the North American canal era unfolded, Upper Canadians and Americans continued to look to each other for ideas, support, markets, technology, and opportunities that would promote a transnational economy, westward settlement, expansion, and improvement in general across their shared boundary.

3

"As Irresistible as the 'Cataract of Niagara'"

The Erie Canal and the Promises of the North American Canal Age

> I wonder how many times the lakes have all /
> Been emptied over here.
> Why Clinton didn't build the Grand Canal /
> From whence I think is queer.
> —Traveler at Niagara Falls.[1]

The above traveler's verse draws attention to the fact that New York State's original Grand Erie Canal which ran from the Hudson River to Lake Erie represented only one of two possible routes envisioned by America's transportation leaders during the nineteenth-century canal age. A popular alternative route to the Erie Canal called for a canal around the American side of Niagara Falls that, in conjunction with a connecting Oswego Canal on Lake Ontario, would overcome the barrier of the Niagara Cataract in the Great Lakes–Atlantic navigation system, and reinforce economic, social, and cultural linkages across the Canada–United States border. However, little attention has been paid to the once popular Lake Ontario route in the history of New York State and the larger transportation revolution. This neglect can be attributed to Erie Canal scholarship that stresses national particularity and Canadian-American commercial rivalry and competition. In the traditional Erie Canal narrative, America's most prominent canal leaders championed the interior Erie Canal route through New York State even though it would be more difficult and expensive to build. They argued that the lake route was too insecurely situated near the British provinces, and that commerce, once afloat on Lake Ontario, would inevitably find its way to the St. Lawrence River and Montreal market. These questions seemed all the more urgent in light of Jefferson's embargoes and the ensuing War of 1812.[2]

There is no question that some transportation leaders favored an interior Erie Canal that would funnel trade across New York State and into the Mohawk–Hudson River. But in the Niagara–Great Lakes borderland region, geographic proximity, family and personal ties, and the mutual need to overcome barriers to transportation and trade mitigated the threat of Canadian commercial rivalry which has long been viewed as paramount in the decision to build the Erie Canal through the interior of New York State. As one scholar observes, Canadians and Americans may have been committed to their political identity and autonomy but "in areas that were contiguous, possessing many common physical characteristics," it was natural that they would mutually share in, and benefit from the chain of navigable rivers and lakes that bordered their two countries.[3] Moreover, as the vision of the Erie Canal unfolded, the question of the competing routes came to revolve as much around local and regional self-interest and development as it did larger questions of ideology and national security. No matter where a person stood on the Erie Canal's route, the importance of the Canadian connection was rarely disputed.

This chapter examines the Erie Canal's conception and construction from the perspective of the Niagara–Great Lakes borderland region. In particular, it considers the role of the largely forgotten Lake Ontario route as a regional and international alternative to the inland Erie Canal, thereby opening to view a significant new dimension of the North American canal era. The contest between these two routes reveals the centrality of cross-border regional trade, cooperation, and development to the shaping of transportation innovations. It provides a counterweight to the conventional focus on broad international rivalries and tensions as the key to the Erie Canal story.

Although nation-centered history has dominated the writing of the Erie Canal, many nineteenth-century transportation visionaries and proponents, especially in the Niagara–Great Lakes region, thought about internal improvements developments in their broader North American context. In 1811 for example, Upper Canada's *Kingston Gazette*, which gave considerable attention to the internal improvements drive south of the border, published a letter by a New York merchant professing that "a more important subject will never be discussed" than that of the Erie Canal in the United States.[4] The letter was in response to the much publicized contest over the projected route of the Erie Canal that began following the passage of an 1808 bill to the New York legislature requesting a survey for the most "eligible and direct route of a canal between the tidewaters of the Hudson and Lake Erie." Marking one of several historic steps in the Erie Canal's progress, Upper Canada's early coverage of the American canal indicated the province's interest and attention to their neighbor's internal improvements schemes. Even during the period of heightening Anglo-American tensions leading up to and following the War of 1812, Upper Canadians closely followed the Erie Canal's progress, suggesting the commercial importance of the American canal to the larger borderland community. Interestingly, one of the chief authors to introduce the bill to the New York Legislature in 1808 was Benjamin Wright of Oneida County, New York, an engineer for the Western Inland Lock Navigation Company whose future service on both the New York and Canadian canal systems would establish him as "North America's premier father of engineering."[5]

Canal enthusiasts along the Niagara–Great Lakes borderland had reason to feel optimistic following the 1808 bill's mandating survey. The bill instructed that both the Lake Ontario–Niagara route and the interior Erie route be surveyed and assessed, which signified the commercial viability of the lake route in the minds of New York State's canal planners.[6] As one authority notes, "the inland navigation corridor between Albany and Oswego, on Lake Ontario, was a highway of commerce and migration" some two decades before the Erie Canal.[7] Moreover, the highly regarded Simeon De Witt, who would hold the post of New York State surveyor general for fifty years until his death in 1834, had expressed his preference for the Lake Ontario route with its call for canals at Niagara Falls and Oswego. An original stockholder in the Western Inland Lock Navigation Company, De Witt favored improvements to the lake route, commenting that the Erie route was "a separate work," to be undertaken only after the Ontario channel was completed. With this vision in mind, De Witt appointed an Onondaga judge and surveyor by the name of James Geddes who, like Benjamin Wright, would come to play an important role on both the American and Canadian canal systems, illuminating the fluidity of the border during the canal age. Niagara's commercial future seemed secure until a local western New York land agent offered the surveyor general more detailed information on an "interior canal" between the Hudson River and Lake Erie that would bypass Lake Ontario and Niagara Falls altogether.[8]

The champion of the interior route was future canal commissioner Joseph Ellicott, agent of the powerful Dutch Holland Land Company, which owned most of the land in western New York, and one of Buffalo's most influential residents. Ellicott, who would no doubt benefit from a canal cut through

Company territory, persuaded De Witt of the advantages of the interior route over that of the Lake Ontario route and its need for a ship canal around Niagara Falls. Ellicott informed the surveyor general that he had devoted much time to the question of a canal along the Lake Ontario–Niagara route, but given the strata and shale around Niagara Falls, believed it too difficult and expensive to excavate. Interestingly, two hundred years later it would take a company more than half a decade, with the world's most modern technology and equipment, to tunnel around Niagara Falls because of the same challenging subterranean forces that Ellicott identified.[9] Meanwhile, surveyor DeWitt responded to Ellicott's letter, remarking that this new information would be taken seriously, and that it might "materially change the idea of the Secretary of the Treasury" on the Ontario route. De Witt's comment was in reference to the freshly penned report on internal improvements by America's longest serving Secretary of the Treasury, Albert Gallatin, whose continental vision of trade and transportation included a canal at Niagara Falls and one at Oswego. The surveyor general felt compelled to note that the interior route would unquestionably bestow "a very great additional value" to the tract of land through which it would pass, a not subtle allusion to the Holland Company's vested interest in the project.[10] In thinking about the most eligible route, the surveyor also raised the sticky question of the recent smuggling activities on Lake Ontario, which came in the wake of Jefferson's embargo restricting trade between Canada and the United States.

Ironically, the surveyor general's remarks about smuggling and the diversion of trade to Canada would have carried the least weight with Ellicott whose transportation and developmental plans for the region included close social and economic ties with the neighboring provinces. Like nearby western New York land agent Charles Williamson, who pushed during the late 1790s for the American Niagara Canal and closer commercial and recreational ties with Canada, Ellicott understood how important the Canadian connection was to the company investment and to the overall health of the borderland economy. Buffalo's settlers naturally looked to the Canadian side of the river for supplies that included hay, flour, tin ware, paper, glass, wines, cheeses, and other sundry products. Many of these goods found their way to Buffalo via the flourishing Niagara portage—a critical transportation and tourist artery that brought an international flavor to the frontier economy and suggested the commercial and recreational potential of the region in the Great Lakes–Atlantic trade network.[11] The establishment of a regular ferry system in the early 1790s that brought people, goods, and information back and forth across the Niagara River affirmed the emergence of a budding cross-border economy in this region.[12]

From the beginning of settlement in the 1790s, Ellicott regarded the Upper Canadian market as critical to Buffalo's future growth and prosperity. In his reports to the Holland Company, Ellicott described the need for transportation routes traveling north so as to better facilitate "communication between our settlement and Chipeway (sic), Queenston, and Niagara in Upper Canada," and provide a general convenience to the settlers.[13] As stated earlier, Upper Canadian merchants welcomed the agent's enterprise and believed their own interests would be greatly benefitted from improvements and settlement on the American side.[14] Even the Holland Company land values were assessed higher along the Niagara River because of the proximity of the lakes and the Upper Canadian market. Ellicott calculated that the commercial prosperity of the Purchase depended on economic links with the neighboring Canadas.[15] The enduring sense of community and friendship that spans the Niagara border today owes its existence to men like Ellicott who early recognized the need to cultivate closer social and economic ties between the neighboring countries (fig.3.1).

Figure 3.1. **Joseph Ellicott.** Holland Land Company agent Joseph Ellicott early encouraged the north-south economic orientation of the Niagara borderland by promoting internal improvements like roads and canals. Courtesy of The Buffalo History Museum. Used with permission.

Though Ellicott supported the interior route of the canal, as a borderland resident and businessman, he understood the commercial and social importance of the Lake Ontario channel to the health of the frontier economy. This became evident between 1808 and 1811 when trade with the northern provinces blossomed. Stimulated by the American embargoes and non-intercourse acts, residents along the Niagara borderland established major smuggling centers, the two most notorious at Black Rock and Lewiston on the American side of the river, which sent illegal produce to the Upper Canadian and Montreal markets. American products and provisions including cattle, wheat, flour, and potash were illegally sold across the border in exchange for some local produce and British manufactured goods. Particularly profitable was potash, a major frontier commodity widely used in the manufacture of soap, which demanded a hefty $300 a ton in the Montreal market—more than three times its usual price.[16] As chief agent of the Holland Land Company, Ellicott felt the positive impact of this illegal trade because cash earned from smuggling helped settlers to finance new land or pay off existing debt to the Company.[17] Encouraged by this wave of commercial activity, Ellicott wrote that should the Lake Ontario route of the canal be chosen "I am persuaded that it will be more advantageous at once to make Montreal our market; and it makes no difference to us what market we go to; the great object is to go to such a place where we can make the most profits, being the governing principles of those who have anything to dispose of." In the same letter, Ellicott observed that "the great men of the nation" favored the Ontario route with its plan for a canal around Niagara Falls, early illuminating both the importance of the lake route and the land agent's interest in further developing this northern corridor.[18]

The Holland Land Company shared Ellicott's enthusiasm for promoting public works programs that would benefit the Company and link the frontier settlement to neighboring Canada. Original shareholders in the Western Inland Lock Navigation Company, the agents of the Holland Company early regarded the Lake Ontario–Niagara route as an important public improvement. During the height of activity surrounding the Niagara Canal project in the 1790s, Theofilus Casanove, the Company's first general manager, requested that Ellicott examine "the practicability and probable expenses of . . . a canal of communication between the waters of Lake Ontario and Lake Erie," as such public works had obvious commercial value to the Company and its frontier settlement.[19] During the next century, Casanove's successor, Paul Busti, similarly contemplated the Lake Ontario route and "[t]he advantages to be derived by the channel of the St. Lawrence to the

North Western parts of the State."[20] Busti came to favor the Lake Ontario route following a series of conversations with David Parish of Ogdensburg, New York, whose profitable mercantile establishments on Lake Ontario and the St. Lawrence River spoke to the advantages of this natural trade corridor. In fact, Busti was so struck by the "highly beneficial" effects of the Lake Ontario route to the western New York settlement that he began to have doubts about the interior canal between Lake Erie and the Hudson. On May 17, 1811, he wrote Ellicott:

> I beg you to devote a page explanatory of the probability or improbability of the Canal of communication with the Hudson being executed. . . . I can perceive but vague plans which leave ample room to doubts . . . I wish you could give some accounts of that interesting business that must materially affect the value of the lands at a distance of even 50 miles from the Canal. Another chapter I beg likewise to devote a minute detail of the benefits the country already derives from the increasing navigation of the Ontario and exports down the St. Lawrence, connected with an enumeration of the establishments already made or about making in the country whose infallible tendency will be to encourage and animate the industry of the agriculturalist in raising products for sale of which they are sure to find a market.[21]

Despite these convictions, the prospect of increased land values in the Purchase eventually prevailed in the land company's decision to support the interior route of the Erie Canal. Still, the beneficial effects of the Lake Ontario channel with its natural links to Canada continued to speak to the sentiments of transportation enthusiasts as they pursued the idea of a canal between the Atlantic and the Upper Great Lakes.

Visitors to Niagara Falls got caught up in the canal route debate. During a tour to Niagara in 1809, fellow scientist and educator Thomas Cooper believed that the commercial prosperity of New York was tied to Lake Ontario and the Montreal market.[22] Traveler John Melish offered a more detailed description of the prospective canals and the centrality of the Lake Ontario-Niagara route in the minds of America's canal leaders:

> The subject of navigable canals having of late excited a great degree of attention in the United States, I resolved at setting out, to pay a little attention to, and have accordingly noticed

it occasionally in the course of this work. From an early investigation of the subject, it appeared to me, that a canal of much importance could be made between this place and Lewiston, so as to continue the chain of communication from the river St. Lawrence along the lakes. There is a natural harbor formed at Fort Schlosser by an eddy in the river, and the ground is pretty level to the extremity of the land above Lewiston. An ample supply of water could be procured from the river, to make up for the waste in the descending locks, so that it is perfectly practicable to make a canal. The distance is about nine miles, and the canal would require to be sufficiently large to admit of sloop navigation.

However, Melish continued that it appears "by a late survey and report of commissioners appointed by the State of New York, that the utility of such a canal may be superseded by a more profitable line to run between the Niagara River and Albany."[23] Melish was referring to the 1810 appointment of the New York State Canal Commission that was assigned to explore the original "Ontario route" from the Hudson River to Oswego, Lake Ontario and around Niagara Falls and the entire "interior route" between the Hudson and Lake Erie.[24] After leaving the Falls and heading back through the Genesee region near Lake Ontario, Melish surmised that the people favored the natural lake Ontario route noting, "The principal market is on the lake, and it is believed by the people here, that it will always continue to be so; they seem so far as I have yet collected their sentiments, to consider the projected canal as of no importance to them."[25]

Despite much scholarly and popular attention to the now familiar interior Erie Canal, the purpose of the 1810 canal commission was not to drum up support for an all-land route through New York State. In fact, a large amount of the commissioners' time was spent analyzing the well-known Lake Ontario channel that served as one of North America's most viable waterways. Indeed, from the beginning of the assignment, the seven-man board of commissioners, the majority of whom were stockholders in the Western Inland Lock Navigation Company which focused during the 1790s on improvements between the Hudson and Lake Ontario, took seriously the responsibility and obligation to examine and consider both routes. Treasury Secretary Albert Gallatin's highly publicized *Report* (1808) on roads and canals brought added attention to the viability of the Lake Ontario channel. In constructing his *Report*, Gallatin looked to the records of the Western Inland Lock Navigation Company to substantiate his faith in this all-important route.[26] Gallatin noted, "A Company incorporated by the State of New York, for the improvement of this navigation, has made considerable success," and continuing this line of improvement to Lake Ontario via a canal at Oswego and another at Niagara Falls was the next logical step. New York Statesmen George Huntington of Oneida County personally supported the Treasurer's plan, stating "a canal by Oswego would unquestionably be the most eligible; besides a canal at Niagara will soon be indispensible (sic); and this must be of sufficient size for vessels that navigate the lakes."[27] All of this information was on hand when the New York State Legislature appointed the canal commission in 1810 to inspect the two channels.

The fifty-three-day-long canal inspection brought the commissioners to upstate New York and Upper Canada where they assessed the agricultural and commercial utility of both routes.[28] Indeed, thanks to the Western Company's improvements to inland navigation, most of the commissioners traveled by canal and natural waterway from Schenectady to Rome, Oneida, Wood Creek, and Oswego, witnessing firsthand the importance of the Lake Ontario channel to both the state and the nation. Having reached Oswego, the commissioners headed next toward Geneva via Seneca Lake before crossing the Ridge Road to Lewiston and Niagara Falls, the hometown of canal commissioner Peter B. Porter. Porter's commercial and recreational interests at Niagara naturally made him a strong champion of the Lake Ontario route. After completing their inspection along the Niagara frontier, and taking a leisurely tour of the Canadian and American sides of the Falls, the commissioners returned to Albany by way of Buffalo along the still contemplated interior route between Lake Erie and the Hudson. Closely monitoring these events in Canada, the *Kingston Gazette* brought attention to the canal commission and its investigation of the much publicized "route for a canal" that one contributor called "praiseworthy" and deserving of Canada's attention.[29]

The Western Company's success in opening an almost continuous line of navigation from the Mohawk River to Lake Ontario allowed the commissioners to travel this route by boat, rather than horse or foot, for much of the journey. De Witt Clinton's private journal, where he dutifully recorded the 1810 commission's work, offers a unique glimpse into the utility of this North American water system some fifteen years before the Erie Canal even opened. During the summer-long inspection, Clinton observed the commercial and recreational activity on this newly improved route, noting such details as the quality of river and lake travel, the variety of boats and river craft, the average carrying capacity of vessels, and the strong commercial trade links with Canada through the Oswego port on Lake Ontario and the St. Lawrence. That a considerable amount of time was devoted to the Lake Ontario route of the canal suggests the utility of this channel and the importance of the Canadian connection in the commissioners' minds.

On July 5, Clinton reported that the commissioners took up their responsibilities at Schenectady, where they boarded a three-ton Durham boat with "a handsome awning, flag flying and large sail," indicating the widespread use of the Durham boat for commercial and pedestrian use at this time.[30] Impressed by the Western Company's improvements to inland navigation, Clinton wrote: "The Inland Lock Navigation Company . . . have five locks at Little Falls, two at the German Flats, and two at Rome, besides their works on Wood Creek."[31] At Little Falls "[t]wo boats passed through the locks in our presence—one a Durham boat" that "draws when fully loaded, 28 inches of water, and can carry one hundred barrels of potash, or 240 of flour." At the German flats, Clinton observed "The canal is 1¼ mile long, 24 feet wide, and 4 feet deep. . . . The lock was filled in five minutes for our boat to pass." The flurry of commercial activity on the canal was observed in the commissioner's comment that during "April and May there passed the Falls, 151 boats." At Rome, the canal "is 1¾ miles long; 32 feet wide at top; from 2½ to 3 feet deep."[32] Meanwhile, Clinton reflected on the aesthetic and recreational appeal of this transportation route while heading toward Oswego.[33] Standing at the juncture of Wood Creek and Oneida Lake, Clinton observed: "To the west the eye was lost in the expanse of waters, there being no limits to the horizon. . . . A number of canoes darting through the lake after fish in dark night, with lighted flambeaux of pine knots fixed on elevated frames, made a very picturesque and pleasing exhibition."[34] All of these improvements to the Lake Ontario route left an impression on the commissioners' minds.

As the commissioners traveled further west along the Mohawk/Oneida/Oswego corridor, they were struck by Canada's role in bringing prosperity and growth to the region, a fact that is often overlooked in nation-centered history on the Erie Canal.[35] At Utica, they learned that much of the town's growth was attributed to the neighboring provinces, it being observed that "[i]n consequence of the trade with Canada, specie is continually accumulating here." Further west, they saw salt boats and lumber rafts "intended to form a junction at Oswego, and to proceed over Lake Ontario, and thence down the St. Lawrence to Quebec. It is supposed they will bring $20,000 at that place."³⁶ According to Clinton, a chief source of business was in the sale of salt to Canada and the American West. In 1809, "28,840 barrels of salt were sent directly to Canada" from Oswego, and "this year it will exceed 30,000." Ninety-ton lake vessels, filled with teas, East India goods, coffees, and other merchandise, also regularly plied between Oswego and the Canadian ports of Kingston, Toronto, and Queenston on the Niagara River.³⁷ A traveler on his way to Niagara in 1810 confirmed the importance of the Upper Canadian market to Oswego, writing, "notwithstanding the embargo," at least "$100,000 worth of wheat, pork, whiskey, and potash" was sent chiefly to Kingston. "The future market of the whole western district of New York" is destined for that market.³⁸ Oswego's residents faithfully subscribed to the Upper Canadian *Guardian*, further illuminating the importance of the Canada–United States connection.³⁹ By the time the commissioners finished their inspection of Oswego, James Geddes had mapped out a canal and locks around the Oswego Falls and rapids. Clinton affirmed Geddes faith in the Oswego project, commenting that "[h]ere a canal might easily be cut around the Falls."⁴⁰

The commissioner's favorable observations regarding trade ties to Canada illuminated the importance of the borderland economy in bringing prosperity to both countries. Though rampant smuggling during the 1807 embargo stimulated some of this cross-border trade, as indicated previously, from the time of settlement in the Lower Great Lakes region economic and personal contacts across the lakes had proven lucrative to residents on both sides of the international boundary line.⁴¹ Moreover, inflated profits made during the Anglo-American conflict further awakened American borderland merchants and businessmen to the profitable Lake Ontario route and trade with Canada.⁴² The strong trade ties between Canadians and Americans in this region made it almost impossible to enforce the embargo or for that matter fight a war between the neighboring countries. If anything, the embargo and non-intercourse acts stimulated trade in this region as more boats were furiously built on Lake Ontario and the St. Lawrence to accommodate the booming illegal trade. In Congress, one New Yorker spoke out against the trade restrictions, noting that his constituents found in Canada a market "for the products of their honest toil, and that their prosperity depended upon this continuing intercourse." Genesee merchant James Wadsworth agreed, noting that "restrictions to trade with Canada embarrass everything. Free trade would be a mutual advantage."⁴³ While tending to his canal responsibilities at Oswego in 1810, Clinton acknowledged that "the embargo enriched the frontier settlements, and the impediments to a free intercourse with Canada" were highly unpopular.[44] In fact, as was the case at Niagara, Oswego had its share of "smuggling agents" who facilitated the illegal trade with Canada. Oswego customs collector, Nathan Sage, drew the line "at hiring smugglers as appraisers" of the illegal goods.[45] Even the looming threat of war could not break the bond between the neighboring countries. Reflecting years later on the war in this region, American President James Madison recalled that "in the portion of the United States connected with the St. Lawrence and the inland seas," there had developed after 1807 "a world of itself "where commercial and personal allegiances defied political boundaries. It was in this world, the president lamented, that America's policies of embargo and non-intercourse had failed.[46]

The story of an early Oswego captain reminds us of the personal and commercial ties across the Canada–United States border that frustrated President Madison during the age of embargoes and War of 1812. Captain Israel Adams, who lived in Liverpool on the northern end of the Oswego Canal, spent much of his life running boats on the inland rivers and Great Lakes and during the Anglo-American war served as a pilot on Lake Ontario and the St. Lawrence River. At the war's immediate conclusion, Adams learned that the people of York (present day Toronto) were in a state of starvation, and made the bold decision to go to Canada, despite the pleas of friends not to go so soon because he might be taken prisoner, even though peace was recently declared.[47] Paying little heed to the warnings, Adams "proceeded with a cargo of Pork, Flour, and Salt, and entered the Bay of Toronto and anchored in front of their batteries, being the first American flag in their harbor after the declaration of peace." Adams recalled that he was wholeheartedly welcomed by "the authorities of the town and commanding officers of the city." After returning from Toronto, Adams returned to the salt business on the lakes and rivers, running a boat between Montreal and Ogdensburg among other interests before retiring at Oswego.[48]

When the canal commissioners reached the Niagara Frontier in the summer of 1810 they were greeted by fanfare on both sides of the international boundary, again underscoring the permeability of the border in this region. Having been apprised of the commissioners' visit, the commanding officer at Fort Niagara welcomed them with "a parade of troops" that, according to one traveler who just happened to be at the Falls on the same day, "were well dressed in every sense of the word, with their arms and accoutrements in excellent order."[49] After dining at the fort, Clinton and a few of the commissioners headed over the Niagara River to Queenston, Upper Canada, where "several of the principal men of the place" came out to see them. Clinton was anxious to hook up with an acquaintance there, a Dr. Robert Kerr who, in addition to being fondly known in the United States as the physician in the British army who treated Americans kindly during the War of 1812, was "Deputy Grand Master of Upper Canada" and a well-respected freemason on both sides of the international border.[50] Interestingly, Dr. Kerr's family was connected through marriage to John Brant, the son of famed Mohawk Chief Joseph Brant, who in later years represented his people's interests when their land on the Grand River was destroyed in part by the Welland Canal Company that put its own interest ahead of the welfare of the native population.[51] After the commissioner's jaunt to Queenston, they crossed back over the river to Lewiston where they lodged that night with Benjamin Barton of Porter, Barton and Company who was also highly invested in the canal outcome.

By the time the canal commission had reached the Niagara frontier, Peter B. Porter had already made known his preference for the Lake Ontario channel in a widely acclaimed speech before Congress in February of that year.[52] As a borderland merchant and entrepreneur, Porter's interest in the Ontario route naturally spoke to a continent-wide canal system that included trade with Canada in the North, the American West, and the Atlantic interests in the East.[53] Gaining the attention of his contemporaries like Albert Gallatin, Porter spoke with eloquence and conviction on the subject of the lake route and a ship canal around Niagara Falls:

> From the place where this canal would connect with Lake Ontario, there is a ship navigation of 200 miles to the Falls of Niagara. A canal, with locks sufficiently large for the vessels that navigate the lakes, might be opened around the Falls, at an expense, estimated by the Secretary of the Treasury at one million dollars. From the Niagara River there is again a ship navigation to every part of Lake Erie . . . And thus . . . a great circumnavigation might be formed, embracing the principal part of the United States and their territories; and connecting in its course, by navigable waters, the whole of the western and Atlantic countries. This canal would open to the navigation of the Atlantic on the lakes above, a coast of between five and six thousand miles.[54]

Armed with Gallatin's "Report," and excited by the commissioner's visit to the Niagara frontier, Porter took the opportunity to promulgate the commercial and recreational advantages of the Niagara ship canal between Lakes Ontario and Erie.

Porter's vast political and economic interests on the American side of the Niagara border factored into his preference for a Niagara canal. In 1805, Porter, in association with his brother Augustus, and Benjamin Barton (the latter who formed the Niagara Genesee Land Company with several British provincials in the 1780s marking it as one of the first borderland ventures in this region) purchased most of the land between Lewiston, Niagara Falls, and Black Rock, as well as other profitable lots along the Niagara River. The following year, Porter, Barton and company attained a lease from the state of New York giving them sole right to transport merchandise across the recently revitalized American portage between Lewiston and Fort Schlosser. The Company also won a federal contract for supplying the western military posts, and quickly expanded its transportation fleet on the Great Lakes. Before long, the Company owned and operated several large shipping and passenger vessels on Lakes Ontario and Erie. Within the next decade or so, the Porter firm came to dominate the transportation, commercial, and manufacturing interests in the Lower Great Lakes region, as well as the newly sprouting tourist industry at Niagara Falls. During the first half of the nineteenth century, the Porter name in the west conjured up similar associations to those of the Livingston's and Van Rensselaer's in the east.

The Porter brother's carrying business on the Great Lakes drew the commissioners' attention to the commercial and recreational interconnectedness of the Canada–United States borderland, and the associated advantages of a ship canal around Niagara Falls. Indeed, no one was more prepared to speak on the advantages of the Lake Ontario route than Congressman Peter Porter whose transportation interests revolved around the Great Lakes trade. In addition to serving Detroit, Oswego, and Ogdensburg, Porter, Barton and Company forwarded merchandise to both the Upper and Lower provinces, as well as operating the ferry service above Niagara Falls that brought people, goods, and tourists back and forth between the two countries. The Porter brothers also had many business contacts in Upper Canada including well-known merchants Thomas Dickson and Thomas Clarke of Queenston, and James Cummings of Chippewa, who regularly contracted with the American portage company.[55] Canada's Welland Canal founder, William Hamilton Merritt, was also well acquainted with the Porter brothers and utilized their company, in the immediate aftermath of the war, to forward shoes, medicines, groceries, and other sundry items from New York.[56] Porter's commercial interests along the Niagara frontier, and his close economic and personal contacts in Upper Canada, naturally led him to favor the Lake Ontario route of the canal.

Porter's position as a renowned national statesmen and future war hero complicated the question of the canal and trade ties to Canada. As one of the region's earliest settlers, Porter daily participated in the borderland economy that had taken root in the 1780s (fig. 3.2). Porter, as observed by one scholar, represented the region of the "northern borderland," a vicinity that "embraced both American and Canadian territories, settlements, and marketplaces along the shores of the Great Lakes and the banks of the St. Lawrence River," and where "the flow and contours of these waterways rather than the boundary lines drawn by the Treaty of Paris in 1783" determined the "movement of ideas, goods, and people throughout this region."[57] Caught between his duties as Congressman in Washington, his responsibilities to

Figure 3.2. **General Peter B. Porter.** A renowned statesmen and War of 1812 hero, Porter was also a borderland entrepreneur and friend of Upper Canada. Courtesy of The Buffalo History Museum. Used with permission.

his portage business at Niagara, and his close relations with neighboring Canada, Porter's position on the canal and ensuing Anglo-American war was complex, yielding conflicting interpretations about his motives and ambitions.

Like most borderlanders during the buildup to and onset of the Anglo-American conflict, Porter would find himself in an untenable position as he tried to reconcile his duties as statesman, borderland entrepreneur, and friend of Upper Canada. Some scholars have argued that Porter's primary reason for taking the position on the canal commission was to secure the Lake Ontario route and his economic and political future. While not denying Porter's attention to his investments, historian Ronald Shaw believed that Porter's "interests centered upon the use of the Ontario route" but that did not impede him from working faithfully and honestly for the commission.[58] Complicating the canal question was the embargo and ensuing war where scholars have charged Porter with everything from profiteering by selling supplies to the army through his portage company at Niagara to that of a "war hawk" bent on expansionist designs and "heedless bellicosity" toward Canada. In contrast, one of Porter's contemporaries defended his good name along the border, noting that the Congressman never put his realty and personal interests at Niagara above "the rights and honor of his country."[59] Moreover, during the buildup to the 1812 imbroglio, Porter, who eventually faced combat along the Niagara frontier, proved reluctant to wage war against neighboring Upper Canada because as he personally stated that he had "many valuable friends in Canada—men with whom he was in habits of almost daily intercourse, and for whom he entertained the highest regard."[60] Only weeks before the war was declared, Porter's chief partner, Benjamin Barton, sent a letter to their business associates and friends in Upper Canada assuring them that they would do everything in their means to avert animosities on both sides, while preparing against any attacks that might take place while "we are doing our business." The respondents acknowledged "that disposition that universally exists here for the preservation of uninterrupted harmony between the two countries" and hoped that the impending crisis would not affect their special relations.[61] The north-south orientation of the border, and the personal and commercial ties of the residents on both sides, was expressed by Porter who lamented that a war with Great Britain "would be peculiarly distressing" and hoped if war was inevitable that it would be acted upon and concluded as quickly and as with as little injury as possible so as to avoid the miseries of a long-contested conflict between the neighboring countries.[62]

Once at Niagara, the canal commissioners turned to the task at hand. Porter and Clinton primarily concentrated on the Lake Ontario–Niagara route, while Morris and Van Rensselaer spent much of their time surveying the still contemplated interior route between Buffalo and Albany. Morris, who brought his new bride along on the canal inspection, may well be one of the earliest honeymooners at Niagara Falls while at the same time establishing the blueprint of the popular nineteenth-century Northern Tour with its theme of canals and transportation innovations as tourist attractions in their own right.[63] Meanwhile, Porter and Clinton followed a map sketched by Geddes showing that the projected Niagara Canal would begin at Fort Schlosser above the Falls and terminate at Lewiston below the Falls, essentially following the route of the portage. The line they followed was familiar and well-documented, having been surveyed as early as 1796 by the Niagara Canal Company. In his journal Clinton wrote: "Gill Creek enters the River on the left bank, about half a mile above Fort Schlosser, and is considered as the probable commencement of a canal. It has a good bay and landing, is deep, and about twenty yards wide."[64] Whether the canal should be made for boats or sloop navigation was debated, but it was the sentiment of Porter, who regularly drew from Gallatin's Report for support that, "in order to be as eminently useful as the nature of the undertaking seems to require," the Niagara Canal "should be on such a scale as to admit vessels which can navigate both lakes."[65] For Porter, Barton and Company which operated sailing vessels of several hundred tons on Lake Ontario and Erie, the Niagara Canal needed to be on a scale large enough to accommodate their commercial interests inland. At the same time, Porter and Company envisioned a ship canal that would facilitate the future growth of steamboat traffic in the Niagara–Great Lakes Basin.

The viability of the Lake Ontario route became evident during the week-long inspection. Taking a personal interest in the project, Porter escorted Clinton up and down the Niagara River where he witnessed firsthand the workings of the Porter shipping empire in this region. At Black Rock, Clinton was so impressed that he wrote: "The rope-walk is sixty fathoms long, is the only establishment of the kind in the western country, and already supplies all the lake navigation." He also "saw wine and jelly glasses here of excellent quality, which were manufactured at Pittsburgh. The common window-glass used here is also brought from that place, and also lead from the mines on the Missouri."[66] The hum of back and forth activity on the Black Rock ferry to Canada reminded Clinton and the commissioners that commercial advantages were connected to transportation linkages with the neighboring country. Also evident was the recreational potential of the projected Niagara canal. In fact, Clinton had the pleasure of staying at Augustus Porter's new residence overlooking Niagara's wondrous Cataract, recalling that "I felt the agitation of the Falls in slightly shaking Judge Porter's house, after I had retired to bed."[67] As the region's leading tourist developers, Augustus and Peter understood that outside of its commercial value, a canal at Niagara would have intrinsic value too. As discussed in a later chapter, visitors just like Clinton would come from afar not only to experience the thrill of the rumbling Falls under their feet but to marvel at the region's most renowned canal innovations like the Lockport Locks and the Welland Deep Cut.

As would become evident in the commissioner's 1811 Report to the legislature, there was much to be said in favor of the lake route with its adjoining canals at Niagara and Oswego. However, later writing downplayed this, arguing that Porter was essentially the only canal member fixated on the Lake Ontario route at this time, and that the only reason he took the position on the Commission was to secure his commercial interests along the Niagara borderland.[68] Yet, the economic, social, and recreational value of the Niagara Canal with its natural trade ties to Canada had invariably left an impression on all of the commissioners. By the time the canal inspection of the Niagara frontier wrapped up, they too had gained a new appreciation of the historic Niagara–Lake Ontario route that Porter and his brother, among others, had faithfully championed.

Just before concluding their summer-long canal inspection, the commissioners stayed at Peter Porter's home in Black Rock, below Buffalo on the Niagara River. It was there that Clinton first observed the *Adams*, "a brig of 150 tons and four guns belonging to the United States" which was "employed in transporting military stores" on the Upper Great Lakes. While inspecting the vessel, Clinton learned that "vessels drawing up to seven feet of water could continue on from the western end of Lake Erie to Chaquagy [Chicago] and then up a creek of that name to the Illinois. . . ." This information gave further credence to the utility of the Niagara Canal that, in conjunction with Lake Ontario and the Oswego connector, would open up several hundred more miles of uninterrupted lake navigation between the far west and the Atlantic. However, the location of the *Adams*, hovering near two British vessels, "one pierced for sixteen and the other for twelve guns," gave ominous warning of the impending war that would ultimately delay any final decisions on the canal.[69] Having finished their responsibilities and taken in the wonder of Niagara Falls one last time, the commissioners returned home by way of the still-contemplated interior route between Buffalo and Albany.

Presented to the legislature in the winter of 1811, the report on inland navigation from the Hudson River to Lake Ontario and Lake Erie has long been regarded as a watershed in the history of New York State's Erie Canal. While recognizing that it would be easier and cheaper to complete a canal to Lake Ontario, the commissioners ruled instead for the "interior" or "Erie" route because they feared that articles "when once afloat on Lake Ontario, will generally speaking go to Montreal." Fear of Canadian competition was said to be acute following Jefferson's embargo, and no amount of legislation or coercion from Washington could prevent the vast smuggling activities that flourished along the Canada–United States border. What was the purpose, then, of cutting a canal from the Hudson River to Lake Ontario, when a canal between Albany and Lake Erie would counter the economic and political risks of the Lake Ontario channel? This question was said to take on greater urgency during the War of 1812 when Great Britain's dominance of Lake Ontario was established.[70]

While it is true that the commissioners deemed the interior route from the Hudson to Lake Erie "practicable" in 1811, the report lacked the definitive or comprehensive plans that many accounts have claimed for it.[71] As *Tacitus* (by De Witt Clinton) later acknowledged: "The short space of time employed in this important affair precluded the idea of an accurate, comprehensive, or definitive report." Whether the interior route "will hereafter be pursued, whether a better may not be found . . . can only be resolved at a future time."[72] It also becomes evident that the question of Canadian commercial rivalry did not dominate the 1811 Report to the Legislature. Instead, technical concerns relating to artificial versus natural channels of navigation, the difficulties of finding qualified lock and canal engineers in America, the exorbitant costs of such undertakings as witnessed by the financially bankrupt Western Canal Company, and the continuing viability of the Lake Ontario channel were all heavily weighted in determining the best route. As more than one study has shown, the question of the canal route was complex with many New Yorkers in 1811 preferring to send their timber and various agricultural products to Canada rather than take the more expensive water route to Albany.[73] The pull of the Canada market and the utility and familiarity of the Lake Ontario route indicated the strength of north-south transportation and trade linkages in the Lower Great Lakes region.

If the 1811 Report ruled on the "practicability" of the Erie route, it also acknowledged that "a good navigation" could be made along the Lake Ontario channel. Indeed, until the opening of the Erie Canal in 1825, and even after, the lake route continued to be the primary outlet for American borderland merchants who sold in the Canadian market. As noted by the commissioners during their tour of the two routes, trade flourished between Canada and the United States in the region of the Great Lakes borderland and efforts to restrict this trade proved unfeasible as indicated by the vast smuggling activities that ensued during the embargo and non-intercourse acts. This was further substantiated by future canal commissioner, Joseph Ellicott, who noted that residents west of the Genesee River found that the Canada market was far more convenient and much less expensive to transport produce to than that of New York.[74] Leading canal authority Ronald Shaw observes that "the factors influencing the selection of a market for the products of western New York were more complex than the commissioner's knew. Montreal was a 'more satisfactory market' by reason of cheaper transportation on Lake Ontario and the St. Lawrence than that offered by the water route to Albany." Produce such as lumber, flour, wheat, pot- and pearl-ashes found a ready market in the neighboring country.[75] As to the question of British trade restrictions on the St. Lawrence, the reality of the borderland and the long history of trade in this region, legal or otherwise, guaranteed that Canadian and American produce would continue to be exchanged on both sides no matter the amount of legislation.[76] Even assuming the canal from Lake Erie to the Hudson was constructed, the 1811 Commission Report recognized the advantage of a side-cut which by means of locks "would connect the canal with Lake Ontario in the harbor of the Genesee," thus opening "a variety of markets" to stimulate trade and "reward the industry of those who are now settled along the Great Lakes."[77]

The publication of the 1811 Report generated much excitement throughout the nation. In January of 1812, the commissioners received a lengthy paper from several Michigan leaders supporting the construction of the Lake Ontario route over that of the Erie. The Michigan leaders acknowledged the importance of a canal from the Hudson to the Great Lakes but argued that the interior route, while operating in New York's favor, offered little advantage to the rest of the nation. The Michigan leaders also spoke of the commercial advantages of the Lake Ontario–Niagara route and trade ties with Canada. The productions of the west, they argued, "[w]hether destined to the St. Lawrence or to the Hudson, whether attracted to Montreal or to New York," must forever pass over Niagara. Once afloat on Lake Ontario, a canal around the rapids of Oswego . . . will present a fair competition between both markets." Commerce will follow the market where the price is highest which "is the only fair and just rule." Competing markets would benefit the economic and psychological health of the nation by offering the producer and consumer a fair and honest price for their commodity. To build a canal solely for the benefit of New York was a "narrow and selfish policy" which sacrificed the interests of the consumer and producer to those of the "mere carrier." In contrast, a canal at Niagara was a work of great utility and magnitude that would "cement the union" and "elevate the national character."[78] The commissioners took seriously the Michigan opposition, noting: "If the Michigan gentlemen were alone in their opinion, it might be useless to say anything, seeing that there is little probability that any contribution will be required of them. However, there are men of influential character who preach the same doctrine. To this effect they assume, what remains to be proved, not only that lock navigation by the falls at Oswego and the cataract of Niagara is practicable, but that it is both cheaper and better" than a canal from the Hudson to Lake Erie.[79] Few men were of more *influential character* than Canal Commissioner Porter who along with his brother Augustus championed the opening of canals from the Hudson to Lake Ontario and around Niagara Falls.

However, the commissioners did have to consider the pleas of several "respectable citizens" from New York City who were increasingly alarmed over the growing popularity of the Lake Ontario–St. Lawrence route and the potential for losing trade to Montreal. During his inspection of the two routes, James Geddes observed that an "interior route"—one that avoided Lake Ontario entirely—would indeed "prevent trade being diverted" to foreign markets, but in terms of the "cheapness of conveyance, the grand desideratum in all such works, it would best be obtained by the Ontario route." These considerations were acknowledged in the Commissioner's 1811 Report, noting: "True it is, that as far as regards the pecuniary benefit of those who may settle along the Lakes, the route by which their products are sent abroad, and the supplies of foreign articles introduced, must be to them a matter of little consequence." Yet, with war looming on the horizon, "the political connection, which would probably result from a commercial connection, certainly deserves the consideration of intelligent men."[80]

While the canal commission continued to promote a canal stretching from the upper Great Lakes to the Atlantic, any serious efforts on behalf of internal improvements were suspended until after the war. In the meantime, the Niagara frontier had become a war zone between the United States and

Great Britain. Still, the commissioners continued to estimate the expense of the canal, made seemingly more urgent by the difficulty of transporting military supplies across New York State to Lake Erie during the inland campaigns. Even as the war raged, it was apparent that the canal still faced challenges and opposition throughout various parts of the state and union. An 1814 canal commission report observed that opposition still existed in the western part of New York where there was a continuing *strong insistence on* the "superiority of what is called the natural communication, by Lake Ontario. . . ."[81] Meanwhile, along the Niagara border, Joseph Ellicott still wavered over the question of the route. Ellicott's interest in the Lake Ontario channel was illuminated during a dinner party in Buffalo that winter, which the land agent attended. Following a ceremonial toast to General Peter B. Porter, whose recent military victories along the Niagara frontier raised his status to that of a national hero, Ellicott raised his glass and proceeded to drink to "the free navigation of the St. Lawrence from its source to the ocean."[82] In addition to revealing the land agents continued interest in the lake route, Ellicott's toast in late 1814 suggested that upstate New Yorkers were anxious to put the war behind them and reestablish commercial, economic, and social ties with their neighbors in Canada. Following the news of the Treaty of Ghent terminating the war, with relief the *Buffalo Gazette* announced that "there have been frequent Balls, in honor of the Pacification. The one of Friday last, held at Pomeroys, Buffalo, was attended by several officers of the British army on the Niagara: and the blue and red coats mixed in the dance with great satisfaction."[83]

The war's end infused the question of the canal's route with a new energy and determination. Perhaps the greatest force in predicting the future outcome of the Erie Canal was not the war and Canadian commercial rivalry, but the influence of several powerful New York City merchants who invited De Witt Clinton to the City Hotel in the winter of 1815 to discuss the city's economic future. In fact, Clinton would use the minutes from this meeting to pen his famous *Memorial* (1816) to the New York Legislature on behalf of an interior canal from the Hudson to Lake Erie. Written at the behest of the New York City merchants, the *Memorial* urged the immediate start of construction on the Erie Canal. The fear "that merchandize from Montreal has been sold to an alarming extent on our borders for fifteen percent below the New York prices," demanded attention.[84] Clinton, who saw an opportunity to bolster his sagging political career after losing his position as New York City mayor, won encouragement by assuring his hosts that the canal would make "New York City the commercial emporium of the world."[85] As Clinton correctly predicted, the Erie Canal would in time make New York City a leading commercial port, and reestablish his political credibility in the state. But for the time being, Clinton was confronted by growing opposition from various quarters of the nation that now disparagingly dubbed the Erie Canal "Clinton's big ditch."

The *Memorial* rehashed some of the earlier themes of the canal commission on inland navigation. The value of canals in providing "cheapness, celerity, certainty, and safety, in the transportation of commodities," and the use of horses in place of manpower, spoke to the "preeminent advantages of canals" over natural river and lake navigation. "A Loaded boat can be towed by one or two horses," resulting in less cost to the seller and buyer. Independent of "winds, tides, and currents," vessels on a canal eliminated "the dangers to which commodities are exposed when conveyed by natural waters."[86] A poem, released a few months after the *Memorial*'s publication titled "The Grand Canal" spoke to the advantages of canals over natural water systems:

> Let Clinton's mental powers unfold
> Who first conceived the project bold,
> To bid the western floods,
> Revolt from nature's long control
> Freely through the hew-mark'd regions roll
> And leave thy astonishing woods.[87]

Of course, just as these convictions were being expressed, steamboats were appearing on North America's inland waterways, freeing vessels from the many natural hazards and concerns of which the poet spoke, and offering fast and reliable transportation for both travelers and merchandize. Indeed, the first steamboat on Lake Erie, the 338-ton *Walk-in-the-Water*, was built by Porter's own Company at Black Rock where it was launched in August of 1818.[88] The *Walk-in-the-Water*, with its advantages of speed and economy, was part of Porter's larger transportation vision which the Niagara ship canal was intended to accommodate. Responding to this major innovation on the Great Lakes, Canada's *Kingston Gazette* wrote: "With all our hearts, we wish her watery walks may be both pleasant and profitable." The same newspaper jested that the skipper of this impressive steamboat, Captain Fish, was "a name better adapted to swimming, than walking in the water!"[89]

Hoping for federal aid, the *Memorial* spoke next to patriotic and national concerns, submerging the obvious local advantages of the Erie

Canal to New York City: "if it be important that the inhabitants of the same country should be bound together by a community of interests, and a reciprocation of benefits," then it is their incumbent duty to support improvements to internal navigation. The appeal to patriotism and the bond of union was particularly directed to American residents along the northern frontier who during the late war continued trading with Canada, despite the economic burden it inflicted on New York City and its citizens. Some Clinton supporters from western New York responded to the appeal by denigrating those who poured into the markets of Canada "the surplus produce of nearly all the western country" and "deprived our citizens of the vast benefits of the western trade." However, the same petitioners had to acknowledge that "men whose object is the accumulation of wealth, will seldom be influenced by patriotic feelings when obedience to such feelings would diminish their expectation of gain."[90] Clinton acknowledged as much in the *Memorial*, recognizing that both routes were feasible and worthy of patronage, and "if the advocates of the route by Lake Ontario did not insist that their scheme should be exclusive, and of course that its adoption should prove fatal to the other project, this question would not exhibit so serious an aspect." That Clinton was open to competing channels of trade was recorded in the next line of the *Memorial* where he says: "If two roads are made, that which is most accommodating will be preferred."[91]

In contemplating a canal for the state, Clinton well understood the economic attraction of the Canadian market to his constituents. The importance of this trade became evident in the immediate aftermath of the war when both countries quickly resumed economic and personal contacts across the border.[92] In June of 1815, for example, a local Upper Canadian businessman advertised in the *Buffalo Gazette* that business "continues at his store near Fort Erie—Dry Goods and Groceries" on hand.[93] During the same time, an ad placed by "J. Wilker, news carrier," informed "the inhabitants of Canada, from Fort Erie to the head of Lake Ontario, that he has made arrangements for supplying them with the news" that could be delivered "at their door steps or a designated place."[94] Commissioner Samuel Street of Niagara, Upper Canada, also advertised in the *Buffalo Gazette* that proposals were being received to build bridges in the neighboring province.[95] That Upper Canadians welcomed any and all measures which facilitated communications and cross-border trade in the war's immediate aftermath was evident in these ads. Recognizing the significance of the Canada connection to the economic and psychological health of the United States, Clinton wrote not long after the *Memorial*'s publication: "The trade carried on between our country and the Canadian provinces is already considerable, and is rapidly growing. The fruits of the earth from the southern shores of Erie and Ontario, and from the borders of the Champlain find their way to the ports of our northern neighbors cheaper than they can to any, which offers a market, of our own, and are there exchanged for the various commodities of foreign countries." This trade "is indeed profitable to many of our citizens who engage in it."[96] While Clinton in no way objected to the Lake Ontario channel, he did argue that the interior Erie route could not be excluded at the expense of the lake one. Only when two equally advantageous paths were opened would the United States be able to compete with the British who, at present benefited the most from trade arrangements on the St. Lawrence and in the Lower Great Lakes.

If Clinton's *Memorial* aroused enthusiasm for his Erie Canal, it in no way stifled the Lake Ontario opposition. Following its publication, Clinton received a letter from Joseph Ellicott who wrote that "from reports now in circulation in this quarter it would seem that the Lake Ontario route is in view and intended to be recommended by his Excellency the Governor in his message to the ensuing Legislature." In a speech delivered to the legislature in the winter of 1816, Daniel Tompkins, who was now one of the most open and keenest advocates of the Lake Ontario route, offered little in the way of support for the Erie channel: "It will rest with the legislature, whether the prospect of connecting the waters of the Hudson with those of the western lakes and of Lake Champlain, is not sufficiently important to demand the appropriation of some parts of the revenues of the state to its accomplishment, without imposing too great a burden upon our constituents."[97] A frustrated Clinton wrote to Ellicott that he thought Peter Porter had forced his opinions on Tompkins who knew nothing on the subject of canals. Porter was also proving inflexible on the question of the route, causing Clinton, whose own political future was tied to the Erie Canal's outcome, to complain to Ellicott "let it not be our fault if it [the canal] is not crowned with success."[98] However, in early 1816, Clinton's canal prospects brightened when Ellicott replaced Porter on the canal commission so that Porter could accept the distinguished federal position of Boundary Commissioner for the United States.

Porter's new position as boundary agent did not weaken support for the Lake Ontario channel. Indeed, so well-known was the lake route to inland commerce, especially in light of the embargo and recent war, that the canal commission in 1816 received a request from the New York City merchants recommending that construction on the middle section of the Erie Canal from Rome to the Seneca River not be delayed because it would have "the most immediate tendency to divert the trade from passing down the Oswego River to Lake Ontario and Montreal."[99] Erie Canal scholar Nathan Miller agrees that it was the New York City merchants who demanded immediate construction on the middle section because "the great commercial emporium at the mouth of the Hudson River" was losing out to the more attractive Canadian market.[100] So strong was the pull of the Canadian market that even the commissioners had to acknowledge "[t]he difficulty of diverting the fixed currents of trade" from this well-known source of trade.[101]

The canal's progress was followed in Upper Canada indicating the importance of the New York canal system to the neighboring province.[102] As already indicated, many Upper Canadians were keen champions of the American internal improvements drive but a few Canadian merchants did express concern that an American canal, by diverting trade away from the Lower Great Lakes and the St. Lawrence, threatened to hurt the prosperity of the province.[103] However, these opinions were not representative of Canadian sentiments in general. As long as Upper Canadians were placed on an equal footing with their American neighbors they welcomed all developments that bolstered trade in this region. In 1816, the *Kingston Gazette* published a letter stating "Citizens of New York" from the western part of the state "continue to speak with confidence of the completion of the Grand Canal from Lake Erie to the Hudson . . . we hope they will go so far as to stimulate the people of this country to improve some of the natural advantages of the St. Lawrence."[104] With both channels free of prohibitive tolls and duties, Upper Canadians had nothing to fear from the American canal. Indeed, as indicated by the province's positive coverage of the Grand Canal, Upper Canadians embraced commercial developments that promoted reciprocal benefits across the border.[105] Canada's positive response to improvements in the Republic affirmed the promises of a transnational borderland economy and transportation network that would benefit both countries.

Clinton and the Erie Canal supporters were determined to push ahead with their project, and when Clinton won the Governor's seat in 1817, a new burst of enthusiasm was infused into the Grand Canal. Porter, a potential rival for the governor's position in 1817 and a leading champion of the Niagara ship canal, might have seen the irony in an *Albany Register* comment that described the Erie Canal and Clinton's renewed popularity to be as "irresistible as the 'cataract of Niagara.'"[106] In a bill passed by the New York Legislature a few months earlier, it was announced that the Grand Erie Canal would go forward. Meanwhile, on July 1, Clinton was sworn in as governor, and three days later on Independence Day, the first spade full of dirt from the Erie Canal was ceremoniously unearthed at Rome, New York, amidst much fanfare and jubilation. Upper Canada's *Kingston Gazette* continued to stay abreast of developments in New York, providing weekly reports of the progress and estimated expense of this "great project" illuminating the province's interest in and admiration of economic developments that promised to benefit both New Yorkers and Upper Canadians in this region."[107]

Yet even in the wake of all this progress, the Erie Canal faced continuing opposition, not the least by Porter and friends who still intended to see their canal built to Lake Ontario. Since the course of the western section of the Erie Canal between the Seneca River and Lake Erie had not yet been decided, the friends of the Erie Canal had reason to be concerned about the Lake Ontario opposition. Though Porter has been accused of undermining Clinton and his Grand Erie Canal for personal and political reasons, in truth far more was riding on the canal's outcome. If the western section of the Erie Canal was abandoned in favor of the Lake Ontario-Niagara channel, Porter's commercial empire on the Niagara River and Great Lakes would be secured. However, should the western section of the canal go to Buffalo and bypass the Falls altogether, everything that Porter and his brother had built along the Niagara frontier might be jeopardized. Moreover, Porter's views on the internal improvements questions were highly regarded throughout the nation. In 1817, a friend of the Erie Canal had to admit that "there are in those waters, gentlemen of talents and respectability who think it will best promote our interests to form all our commercial connections with Canada."[108] Meanwhile, during the same time, a motion to build a canal around Oswego Falls was raised in the Senate, indicating continuing widespread interest in the Lake Ontario channel. Defeated by only one vote, the primary opposition to the Oswego Canal came from New York City.[109] Several months later, Charles Haines, a close friend of Clinton and loud proponent of the Erie Canal, spoke about the power of the Lake Ontario opposition, noting "that a work of 400 pages is in now in the press, to prove that the western canal should be abandoned, for a canal around Niagara Falls."[110]

But Clinton's reelection as Governor in 1820 put him in a strong position to push the canal forward. The opening of navigation on the middle section of the canal that October further united people's energies around the project, and before long, contracts were being let out for the eastern and western sections.[111] Meanwhile, Peter Porter and friends, having lost the first battle over the Erie Canal route, began a new campaign to bring the far western terminus of the canal to Porter's village of Black Rock instead of Buffalo at the foot of Lake Erie. Despite these efforts, Buffalo eventually won the prize of the much coveted western terminus. Clinton's vision of a Grand Canal stretching all the way from the Hudson River to Lake Erie was finally coming to fruition.

From the inception of the Erie Canal to its celebratory completion, many Upper Canadians welcomed the anticipated commercial and recreational benefits of the American project. An 1818 letter in the *Kingston Gazette* noted that "transportation from Albany, on the Hudson to Buffalo on Lake Erie, will not exceed 25 cents per hundred," indicating the advantages of the canal to people engaged in trade on both sides of the border.[112] The following year when the middle section of the Erie Canal was open to business, the *Kingston Chronicle* wrote "boats are already floating" on parts of the "The Great Western Canal" which "the public will now be enabled to appreciate . . . this incalculably important project."[113] A Kingston citizen added: "For my part, I rejoice to see the energies of our active neighbors, proceeding on the basis of peaceful arrangements. I can perceive no reason for lamenting the advantages which New York promises herself from the completion of this stupendous undertaking. She deserves every benefit and every encouragement for attempting so grand a work." The same individual continued that the "canal presents an easy and cheap conveyance to New York whenever that market is better than Montreal" and further asserted that "this canal will be as beneficial to Montreal as it will be to New York, providing the transport be as cheap after leaving the canal, as by continuing on it." The "true basis of commercial intercourse is reciprocal benefits" concluded the Kingston native. This feeling of good will came only three years after the war's conclusion signifying the canal's role in eliminating contention between the two peoples.[114]

In addition to reciprocal trade benefits, advantages also came to Canadian laborers who found work on the Erie Canal. As mentioned in an earlier chapter, Canadian laborers contributed to the improvement of the Mohawk River in the 1790s under the supervision of the Western Inland Lock Navigation Company. In 1823, the *Kingston Chronicle* commented: "We hear that much is the rage for going to work on the New York Erie Canal, that servants, either male or female can hardly be had at York at any price. Outdoor servants demand nine and ten dollars per month, and women servants seven and eight dollars, and even at these extravagant wages, few are to be had."[115] As well as revealing the presence of female workers on the American canal, such accounts indicate the free movement of Canadian laborers across the international border as they went in search of jobs and other opportunities in the neighboring Republic.[116] As indicated earlier, Canadian laborers continued to seek employment opportunities on the American line, it being announced by one Canadian newspaper that "three thousand" men were "on the great canal between Lake Erie and the Hudson. Among the laborers thus employed, a considerable number are said to be emigrants who arrived at Quebec from Europe."[117] On the heels of the Erie Canal's completion, Canadian laborers also found employment opportunities on the Oswego Canal and other American projects.[118]

Many Upper Canadians rejoiced in the Erie Canal's progress. As early as 1819, the *Kingston Chronicle* announced that "the Governor of the State of New York, together with the commissioners, made a trial of that part of the middle section of the Erie Canal which has been completed . . . the boat was 61 feet long . . . and towed by a single horse with apparent ease."[119] In the following weeks, Upper Canadian newspapers recounted the Governor's historic trip between Rome and Utica on the Erie Canal as a courtesy to their readers, and thereafter provided regular coverage of the 363-mile-long canal as it cut its way to Buffalo.[120] "As everything relating to the American canals cannot fail to excite the interest of the Canadian population," wrote the *Kingston Chronicle*, "we copy into our columns the following article relating to the Erie Canal." The *Chronicle*'s repeated coverage of the American canal suggested that it was not only New York City merchants who would benefit from the immediate opening of the middle section, but that Upper Canadians also hoped to prosper from the new outlet.[121] Clinton and his "magnificent projects," wrote the *Kingston Chronicle*, particularly the Grand Canal, "cannot fail to raise the state of New York to the highest pinnacle of prosperity."[122] As the Erie neared completion, the *Upper Canada Herald* informed its readers that "the Great Western Canal" is almost finished "except about 30 miles at the western end from Lockport to Buffalo."[123] Not long after, the *Niagara Gleaner* informed the public that the American canal "is the most economical means of travelling now known."[124]

But perhaps the most important indication of the strengthening cross-border ties was Upper Canada's response to the historic *Wedding of the Waters*, which officially commemorated the Erie Canal's opening in 1825. On the morning of October 26, the day of the gala event, residents along the Niagara borderland were in a tizzy preparing for the much publicized occasion. At precisely 10 a.m. on the morning of October 26, Governor De Witt Clinton boarded the *Seneca Chief* that lay waiting in the Buffalo harbor to begin the much publicized canal parade to Manhattan. Several committee members, engineers, and invited guests rode with Clinton on the *Seneca Chief*.[125] Sadly missing among the list of dignitaries was Joseph Ellicott who, though he had contributed much time, and had arranged for the transfer of more than 100,000 acres of land to the Erie Canal project, was wasting away in a New York City asylum after a life-long affliction with depression.[126] General Peter B. Porter, who by now was regarded as one of the nation's most important statesmen and War of 1812 heroes, chose not to accompany Clinton and the entourage. Though Porter's absence may have been connected to his old rivalry with Clinton, canal historian Ronald Shaw writes that the General was probably just trying to avoid all of the speeches and eulogies that were required of the distinguished crew as they wound their way from Buffalo to Albany and so opted to travel alone.[127] But just as the *Seneca Chief* entered the Buffalo harbor amidst a battery of gun salutes and exuberant well-wishers, Porter was spotted in the canal parade, sailing on board his own vessel fittingly called the *Niagara* that pulled ahead of the flotilla and arrived in Manhattan three days early. Apparently, the *Niagara* was "the first vessel that made the entire passage of the canal between Buffalo and New York City."[128]

The *Wedding of the Waters* ceremony that culminated in Governor Clinton's historic pouring of a keg of Lake Erie water into the Atlantic Ocean at Sandy Hook was celebrated on both sides of the international border. During the Buffalo festivities, crowds of well-wishers came out to witness the celebration, and sounds of rejoicing echoed across the Niagara Frontier. Upper Canadian newspapers, including the *Kingston Chronicle* and the *Upper Canada Herald*, provided coverage of the "Celebration of the Completion of the Erie Canal."[129] Several days later when Clinton's flotilla arrived in New York City, the *New York Mirror* reported with satisfaction that there were two British sloops at the canal celebration, the *Kingfisher* and *Swallow*, which "fired salutes on the approach of the flotilla" and that the "gratifying compliment was returned by all the steamboats making a circuit around the sloops of war and giving them three cheers." The captain of the *Swallow* served an elegant breakfast to a numerous company of ladies and gentlemen and "emblematic devices" like "the American eagle on one side and on the other the British crowned lion" were tastefully displayed on the table. This symbolic gesture on the part of the *Swallow*'s captain toward his American guests indicated the role of canals and other internal improvements in bringing goodwill and friendship between the neighboring countries.[130] Picking up on this important story, the *Buffalo Emporium* cheerfully reported that the British sloop "fired a national salute of 24 guns" and even had "the American ensign displayed at the foretop as a mark of respect."[131]

The commercial benefits of the Erie Canal quickly bore fruit in the Niagara–Great Lakes Basin. Swarms of canal boats arrived from the east carrying emigrants headed for Upper Canada and the American West, alongside goods and tourists. Local newspaper accounts provide some insight into the growing cross-border trade. Anticipating the Erie Canal's advantages, the Canadian *Niagara Gleaner* wrote "It can hardly be doubted that, in peace, a trade will be carried on between the United States and Canada. The termination of the canal being immediately on the frontier, the canal boat can cross the river, and either deliver such articles as may be intended for that country, or bring in the articles that may be intended for Canada."[132] The canal had hardly been completed when it was announced that the *first boat* at Buffalo had arrived and "was freighted with oysters" and other fresh produce "in confirmation of the great event which has united that distant region with the Atlantic."[133] Previously, freight could take weeks to travel from New York to the Niagara frontier, but now produce like fresh oysters, salmon, and clams were being received in impressive quantities. Monitoring this development, a local Upper Canadian newspaper enthusiastically announced that commercial opportunities were to be had in shipping "Fresh Salmon from Lake Ontario" to the New York market "via the Erie Canal."[134] For one Buffalo tourist, the continuous arrival and departure of canal boats and vessels on the lake gave the region "the appearance of a seaport, while the freight of the lake has so increased that it is doubtful whether the vessels employed will be able to carry it before winter sets in."[135] Indeed, as shown in the next chapter, the boom created by the Erie Canal in this region created a greater demand for the construction of the adjoining Welland Canal that promised to accommodate the overflow of traffic from the Erie. "The economic logic of the Erie system," writes one scholar, "promoted the integration of Upper Canada with the economy of the northeastern . . . United States."[136] In the long run, scholar Gilbert Tucker acknowledged that "the opening of the Erie to their trade did more good than harm to the people of Upper Canada."[137]

For the friends of the Erie Canal on both sides of the border, mutual advantages came from cooperating with the opposite side. Advertisements placed in local American newspapers reflect the perceived importance of the Erie Canal in raising land values on the Canadian side of the Niagara borderland (fig. 3.3). In 1826, William Powell of Bertie, Upper Canada, announced that he had for sale 260 acres of excellent land under cultivation on the Niagara River that "lie nearly opposite the Erie Canal, and will be the landing place of the ferry boats from the north end of Squaw Island to Canada."[138] The following year, William Smith of Waterloo, Upper Canada advertised his Farm for sale "at Waterloo, opposite and within one mile of Black Rock, containing about 170 acres . . . under improvement and good cultivation." On the premises were "a commodious house . . . a barn, stable, ice-house, garden, a young orchard bearing choice fruit." The land's value was enhanced by a view of the "Erie Canal, Harbour and basin of Black Rock and the Lake."[139] Such advertisements suggest the Erie Canal's positive role in stimulating economic and commercial activity in the Niagara region.

Merritt's biographer, who took a keen interest in everything connected to the Welland Canal, wrote that the excitement surrounding the Erie Canal

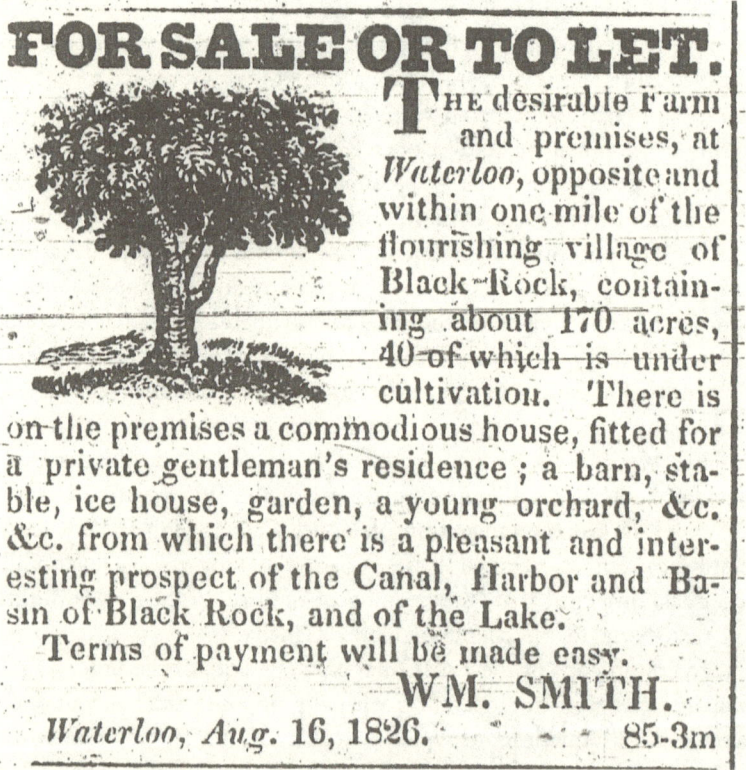

Figure 3.3. **The Canal Age at Niagara.** As indicated by this illustration, the Erie Canal created new opportunities along the Niagara borderland.

created "a mania for canalling" along the Niagara borderland. Advertisements placed in the neighboring province by American businessmen now regularly announced that "Erie Canal" packet boats were offering competitive fares for people traveling from Albany to Buffalo, hailing in Upper Canada the commencement of navigation on this new and important line.[140] Thanks to the American canal, wrote Canada's *Niagara Gleaner,* a year of so earlier, a "travelling mania" is upon us.[141] Discussing the success of the Erie Canal and "the magnificent forbearance of De Witt Clinton," Upper Canada's *Farmer's Journal and Welland Canal Intelligencer* in 1826 acknowledged: "From the experience of the people of the United States, Canada may derive many permanent advantages."[142] Recognizing the popularity of packet travel unleashed by the Erie Canal, several businessmen convened in St. Catharines, Upper Canada, the home of the future Welland Canal, to discuss the expediency of "building a Packet Boat" to run on the Canadian line.[143] Within a few short years, a line of packets and freights were regularly running between Port Robinson on the Welland Canal and Buffalo at the western terminus of the Erie Canal.[144] Packets, freighters, locks, and line boats became new conversation pieces as canal travel and thought absorbed the attention of residents on both sides of the border. The Erie Canal introduced a new and more comfortable mode of business and pleasure travel not only to the first canal generation of New Yorkers, but to Upper Canadians as well.[145]

The Erie Canal was greeted by many in both the United States and Upper Canada with anticipation and wonder. As already observed, Upper Canadians regularly applauded the developments and enterprise of their American neighbors, particularly the genius and industry of De Witt Clinton who was considered "the best Governor the state of New York could ever have."[146] In fact, a funny joke about Clinton's popularity on both sides of the border was recounted in a Buffalo newspaper a year after the Erie Canal's celebratory opening. According to the joke, a Clinton supporter walked up to a gentleman on the street and politely said "he did not wish to intrude, but would be glad to know if Mr. Clinton was elected, and by what majority?" The gentleman answered that "Mr. Clinton was doubtless elected, but that the majority was uncertain, as the returns from Ohio, Michigan, and Upper Canada, had not as yet come in!"[147] Upper Canadians closely followed the election of their favorite American statesman De Witt Clinton with the *Upper Canada Herald* going so far to claim that Clinton would have lost the election if not for "the great undertaking" of the Erie Canal.[148]

The popularity of De Witt Clinton and his Grand Canal in Upper Canada was made evident in 1824 during a dinner party at Albany attended by the Canada Land Company agent John Galt, and Governor Clinton and his wife. The Canada Company provided for the settlement of western Upper Canada which Galt, as one of the company's chief agents in North America, oversaw. After visiting with the Clintons, Galt traveled the Erie Canal to Buffalo affirming his belief that the American channel would bring European emigrants to Upper Canada much more cheaply and efficiently than the St. Lawrence River. While some British officials worried that the American canal might re-route the emigrants to the fertile United States, Galt, who was now highly impressed with both the direct Hudson–Lake Erie route, and Governor and Lady Clinton, pushed for the stationing of company agents throughout North America, including the ports of New York, Buffalo, and Oswego. Because of Galt's efforts, company agents, along with sailing times and passenger and cargo fares, where regularly posted to accommodate the waves of emigrants headed for the upper province. The Canada Company advertised that "vessels will be in readiness" at the port of Buffalo for those emigrants heading to Upper Canada. "Settlers will be transported from this place with their baggage at the expense of the Company."[149] Even after the improvement of the Welland–St. Lawrence line in the late 1820s and 1830s, the Erie Canal system continued to serve the interests of the Canada Company, as well as facilitating Canadian settlement.[150] The role of internal improvements in promoting cross-border economic and social linkages in this region was becoming increasingly evident.

Having successfully launched his grand Erie Canal, De Witt Clinton could now speak more assuredly of other internal improvements schemes, offering encouragement and support to sister canal projects in both the United States and Canada. Clinton was impressed by the energy and importance of the internal improvements drive in both countries, championing the construction of better "[c]ommunications between the waters of Lake Ontario and the western canal, between the River St. Lawrence and Lake Champlain, and between the Susquehanna River and the Seneca Lake."[151] Indeed, during the much heralded "Wedding of the Waters," Clinton and the celebrants made a special toast to the Oswego Canal that was hailed as "an important link in the chain of communication."[152] As Clinton's close friend and biographer, David Hosack wrote, Clinton's plans for internal improvements "were not subscribed by geographical limits or even by national policy," and his generosity and interest in Canada's Welland Canal "was highly reflective of Clinton's countenance and faith toward all such important works."[153]

Following the historic "Wedding of the Waters," Clinton's companion during the 1810 canal inspection, Peter B. Porter, found ways, along with his brother Augustus, to benefit from the canal mania. Peter returned to his hometown at Black Rock where he busied himself with his mercantile pursuits, his new responsibilities as international boundary agent, and the still unfulfilled dream of a ship canal around Niagara Falls. Peter and Augustus began to expand their commercial and recreational interests along the Niagara borderland by constructing steamboats on Lake Erie and the Niagara River and barges for use on the newly opened Erie Canal. In fact, a story is told that during the commemorative Erie Canal opening in 1825, Augustus Porter sent the first shipment of apples east along the canal to Troy and New York City, illuminating the early beginnings of the fruit trade in the Niagara region.[154] Meanwhile, with the Erie Canal completed, and developments like the soon-to-open Oswego and neighboring Welland Canals moving forward, the Porter brothers continued focusing their business interests on the complementary role of transportation and tourism in bringing growth and prosperity to the Niagara–Great Lakes region.

In 1826, now famed Erie Canal engineer Nathan Roberts was appointed to once again analyze and survey the prospective utility of a canal around the American side of Niagara Falls. Roberts was one of the most sought-after engineers in the United States and Canada because of his ingenious creation of a five-tiered set of locks at Lockport at the western end of the Erie Canal.[155] Spurred in part by the Erie's success, and the rapid strides taking place on the neighboring Welland Canal, Roberts believed that if the United States was to remain competitive, the American Niagara Canal needed to be "proportioned to the largest class of steamboats and schooners navigating the lakes, and correspond with the ship or steamboat canal on the St. Lawrence" so that "vessels from the ocean can be passed to our upper lakes." But rather than threatening American interests, Roberts was confident that improvements like the Niagara ship canal and those going forward on the Welland Canal and the St. Lawrence River would make for "formidable but noble rivalries."[156] American journalist Hezekiah Niles of the *Niles Weekly Register* agreed that while it was important "to secure to our country every advantage that it should rightfully possess, we earnestly wish the opening of more avenues to the seaboard from the interior." In fact, Niles believed that with a ship canal on the American side of the Falls, in conjunction with the Erie Canal and soon to open Oswego Canal, "we should have little to fear from competition," but "even be aided by the Welland Canal."[157] However, as indicated in the following pages, the fate of the Niagara canal was sealed in 1827 when the Welland Canal, at the request of the New York investors, was expanded from a barge to a ship canal, eliminating any real future prospects for an American ship canal. By this time, too, Roberts, and several leading American engineers and builders were already busy giving advice and assisting on the neighboring Canadian channel. Canals and other internal improvements were overcoming both artificial and natural barriers in the Niagara–Great Lakes Basin, while also shaping a cohesiveness and connectivity across the border independent of the international boundary line.

4

"A Salutary and Desirable Competition"

New York State Influence in the Building of the First Welland Canal

> When this canal [the Welland] is finished, vessels will pass on its waters, carrying 160–200 tons; and it will open a free and uninterrupted navigation, from Chicago, at the foot of Lake Michigan, its whole extent, and also through Lakes Huron, St. Clair, and Erie, to Oswego on Lake Ontario; from whence to Syracuse on the Erie Canal. A canal is now cutting [the Oswego], and is nearly completed, connecting Lake Ontario with the Erie Canal.
>
> —A Niagara Traveler (1826)[1]

Historians have largely analyzed the First Welland Canal (1824–1829)[2] in terms of national competition and rivalry in relationship to America's Erie Canal. The story is told that in their bid to redirect trade away from the increasingly competitive American channel, Canadian and British strategists furiously built the Welland Canal—a major first step in the improvement of the Great Lakes–St. Lawrence water system, and the development of a national market that would unite the far-flung Upper province with the eastern seaboard interests.[3] However, while the more distant colonial elite may have viewed the Welland Canal as a defensive measure against American transportation developments, along the Niagara–Great Lakes borderland the story of the Welland Canal's building was one of Canadian-American cooperation and friendship. Largely planned around the Erie Canal's 1825 completion, the Welland Canal drew on American finance, labor, engineers, and technology that was now available to assists public works efforts in Upper Canada. Encouraging the development of the Upper Canadian canal system in untold ways, the United States both contributed to, and benefitted from the Welland Canal (particularly following the opening of the Oswego channel),[4] illuminating the point that lines of transportation and trade ran north and south, as well as east and west, and that common interests and visions often transcended national boundaries in the Niagara–Great Lakes Basin.[5] As discussed in this chapter, the complementary nature of the canal age in this porous region strengthened commercial, social, and cultural ties across the international border while providing local benefits and opportunities to borderland merchants, businessmen, developers, and canal leaders whose day-to-day business and personal activities and interests were increasingly tied to the interconnected and interlocking water system.

Just as New York Governor De Witt Clinton is given primary credit for advancing the Erie Canal, so Upper Canadian politician and transportation leader William Hamilton Merritt is recognized as the chief inspiration behind the Welland Canal (fig. 4.1). Born in Westchester County, New York, in 1793, Merritt was the son of a late loyalist who migrated to the Niagara district in Upper Canada when William was only three years old. Merritt's public career in the upper province began during the War of 1812 where he earned the rank of captain before being captured by the American army at the Battle of Lundy's Lane.[6] Held prisoner in the United States until the war's end, Merritt's ordeal was eased somewhat by family and personal contacts on the American side of the river. Following his capture at Lundy's Lane, Merritt wrote in his journal: "[W]e were marched to Major Miller's Tavern, two miles distant from Buffalo, where we had good fare; I saw my uncle William . . . who was then residing here. Colonel Chapin was very attentive."[7] For the entire span of the war, and even while being held prisoner, Merritt regularly corresponded with his American fiancée Catharine Prendergast, the daughter of Dr. J. Prendergast, a western New York State Senator[8] who earlier practiced medicine in Upper Canada before returning to the United States.[9] The rich correspondence between Merritt and his future wife indicates how the border remained permeable despite the war that divided their respective countries. In fact, during his imprisonment, Merritt confessed in a letter to Catharine that he "secretly wished" to remain captive in the United States so as to remain closer to her.[10] Even while fighting for the British, Merritt was reminded of his dual heritage and

Figure 4.1. **William Hamilton Merritt.** Statesman and Father of the Welland Canal, William Merritt was often affectionately dubbed the "De Witt Clinton of Canada." J. P. Merritt, *Biography of the Honorable W. H. Merritt*. Image courtesy of the University of Alberta.

continuing dependence on the United States for his social, economic, and political well-being, a relationship that would evolve more concretely during the unleashing of the transportation revolution in the war's aftermath.

Given Merritt's personal and familial ties to the United States, and the internal improvements fever that followed on the heels of the war, it was not surprising that he should look across the border for assistance and encouragement in the building of the Welland Canal. In fact, Merritt's father-in-law took a keen interest in canals and other internal improvements and counted De Witt Clinton among his many friends. Regarded as one of western New York's most influential citizens, Prendergast was a proponent of both the Erie Canal as well as his son-in-law's future canal venture in Upper Canada.[11] Though Merritt lived out his life in Upper Canada, he regularly visited New York on canal business or vacationed with his in-laws on Chautauqua Lake. When not in the United States, Merritt kept abreast of American commercial and political affairs by regularly subscribing to a New York newspaper, indicating his close relationship and attachment to the neighboring country and the spread of ideas and information across the Canada–United States border.[12]

If the internal improvements drive in neighboring New York would awaken Merritt to the commercial importance of a canal between Lakes Ontario and Erie, it was the prospect of improving his property along the Twelve Mile Creek in Upper Canada, and not the threat of national competition and rivalry with the United States, that provided the initial stimulus to push the canal forward.[13] On returning from the war, William and his American bride settled in St. Catharines, a small village in the Niagara region that would become home to the Welland Canal. Over the next few years, Merritt pioneered an impressive milling and salt industry on the Twelve Mile Creek near his residence, and it was out of the need for water to power his mills that the idea of a canal grew. However, before long Merritt's plan for an irrigation ditch to remedy a water shortage on his property grew into a much larger plan for a canal carrying boats. In 1818, with the backing of several local businessmen, Merritt petitioned the legislature for support in creating a line of communication between Lakes Ontario and Erie by means of the Twelve Mile Creek and the Welland River, but the high cost of the undertaking, and the scarcity of money following the war, diminished the hope of government involvement in the project.[14]

Though the Welland Canal would eventually be recognized throughout North America as the most important in a series of Great Lakes–St. Lawrence canal projects, it would take Merritt another five years to garner enough support to push the canal forward. Frustrated by the lack of enthusiasm for his canal, in 1818 Merritt wrote to his father-in-law in New York, "if your people had it it would be accomplished in one year, and the benefits from it would be incalculable."[15] However, as the project grew from an irrigation ditch on his property to a ship canal carrying boats around Niagara Falls, few doubted the logic of this visionary work. Until the Welland Canal, ships sailed from the head of the St. Lawrence to Lake Ontario and through to Lakes Erie, Huron, and Michigan uninterrupted save for the barrier of Niagara Falls. By traversing the Niagara peninsula and bypassing the Falls to connect Lakes Ontario and Erie, the Welland opened hundreds of more miles of uninterrupted navigation between the continental interior and the Atlantic. It was this foresight that would lead many to dub Merritt the Clinton of Canada.[16] Merritt, who spent a life time admiring and imitating the energy and enterprise of the neighboring Republic, would have been flattered by the comparison to America's father of the Grand Erie Canal.[17]

In the spring of 1823, Merritt had a survey made of his projected canal route. The question of the canal's feasibility was entrusted to Hiram Tibbett, an American engineer off the Erie Canal who conducted the first survey on the Canadian project.[18] Representing one of many American engineers who would cross the border to assist public works in Upper Canada, Tibbett confirmed Merritt's view: a canal could be cut across the Niagara peninsula from Lake Ontario and Lake Erie by way of the Twelve Mile Creek and the Welland River.[19] Wasting no time, Merritt and several of his enterprising friends petitioned the legislature for permission to build the canal, and in January of 1824 the private Welland Canal Company was created.[20] The perception that the Welland Canal would serve as a defense against American commercial and military aggression may have held stock among the more distant colonial elite, but along the Niagara borderland, the Canadian project was largely regarded as a complement to the Canada-United States canal age that was just getting underway in this region. The *Niagara Gleaner* spoke to this sentiment, writing:

> We are glad to hear it held out as probable that a trade between the ports of the United States on Lakes Erie and Ontario is contemplated through the Welland Canal—this looks better than the forbidding propositions of having a canal as far away from the frontiers of the United States. While we are friends let no jealousies exist, but let us in our commercial pursuits, go hand in hand.[21]

As Merritt and friends contemplated the Welland Canal, America's proximity along the frontier continued to be perceived not as a liability, but as an asset that would promote greater commerce and growth between the contingent countries.

Upper Canada looked to New York State and the Erie Canal as a model for their undertaking. In a widely distributed circular, Welland Canal President George Keefer wrote "The President and directors have determined on appealing to the public spirit of the landholders throughout the province—A similar measure was adopted in the State of New York prior to the commencement of the Grand Erie Canal, and donations, in land, received to the amount of *one million dollars*." Keefer was referring to the public spiritedness of the powerful Holland Land Company of western New York that had donated more than one hundred thousand acres of land to the state, and other generous Americans like Gideon Granger of Canandaigua, New York, who gave more than 10,000 acres of land to move the American project forward. Keefer observed a zealousness and commitment toward internal improvements in the neighboring Republic that was worthy of imitation in Canada.[22] Canada's *Colonial Advocate* shared this view commenting "there is not a canal in the United States but what yields, or bids fair to yield, a handsome profit."[23] Like Keefer, and his close associate Merritt, many Canadians admired the Erie Canal and its founder De Witt Clinton who epitomized the American entrepreneurial spirit. As shown earlier, Clinton was himself a friend of the Welland Canal, going so far as to comment to Merritt: "[Y]ou have physical advantages on your side, but you want men of enterprise like yourself, to carry them through."[24] So admired was Clinton in Upper Canada, that upon hearing of his untimely death in 1828, the *Farmer's Journal and Welland Canal Intelligencer* wrote "his genius was his own—grand and peculiar," and his death has left not only New Yorkers but Upper Canadians "deeply stricken with a sense of great loss." (See fig. 4.2.) In New York City, the British consulate requested that "the masters of the several British ships now in port, upon Sunday next, to hoist their colours at half-mast, as a mark of respect for the memory of his Excellency De Witt Clinton"[25]

The United States proved an important source of financial support and encouragement in the Welland undertaking. Thanks to his father-in-law, who had a great deal of influence in Albany, Merritt was introduced to New York's wealthiest businessmen and canal investors. Promoting the Canadian canal while on business in New York City, an exuberant Merritt wrote to his wife: "I have succeeded far beyond my most sanguine expectations. Have got the necessary amount of stock subscribed by the most respectable and influential men in the money market of New York."[26] With much interest did the *New York Albion* note that fifty thousand dollars' worth of stock had already been taken by American investors in the Canadian company.[27] The Company president found a similar interest among American investors, commenting "I accepted subscriptions to the amount of $75,000, and might, I think, have got the whole amount if it had been wished for."[28] The entire stock in the company would have been bought up by American investors but some was reserved for Canada and the European market. The *New York Daily Advertiser* wrote that a number of Americans "were very anxious to have the work completed; and that all the stock would have been readily

Figure 4.2. **De Witt Clinton.** The leading force behind the Erie Canal, Clinton's untimely death in 1828 was mourned on both sides of the Canada–United States border. Courtesy of New York Public Library Digital Collection.

subscribed for in New York."[29] Considerable stock in the Welland Canal Company was also held by neighboring Buffalonians providing further evidence that nationalist sentiments did not keep New Yorkers from investing in and pushing forward the Canadian project.[30] Indeed, just as the United States helped finance the Welland Canal, so Canadians too invested funds in the New York canal system.[31]

Though New York State would come to control the largest number of private shares in the Welland Canal Company, Merritt did not look solely to the United States for support in his undertaking. The magnitude of the canal required vast reserves of capital and just as New York State found it necessary to finance the Erie Canal by looking to foreign investors, so Merritt would travel to England in search of willing sponsors.[32] The list of British investors in the Welland Canal was quite distinguished, including the Duke of Wellington, of Waterloo fame; William Huskisson, a leading British reformer and friend of the transportation revolution who was ironically killed by a railroad locomotive in 1830; Alexander Baring, of the world-renowned Baring house of finance which invested in the Erie Canal; and Edward Ellice, a fur trade baron whose family invested fortunes in land and mercantile pursuits in Canada and New York.[33] Dozens of lesser known Canadians held small amounts of stock in the Company, at least three of whom were women, indicating the latter's role in the transportation revolution.[34] However, it was to New York State that Merritt came to feel most keenly indebted. Without American backing and support, the Welland Canal would have likely gone the way of many failed private canal companies in North America. The total cost of building the Welland Canal would eventually amount to just over four hundred and fifty thousand British pounds.[35]

Among the American investors, John B. Yates of Chittenango, New York, became the single largest shareholder in the Welland Canal Company.[36] As will be seen, Yates had a long history and affinity with the Upper province, and by his death in 1836 would take much credit for supporting and promoting the internal improvements drive in the neighboring country. Born in Schenectady, New York, in 1784, he served with distinction during the War of 1812, sat as a Republican in the prominent Fourteenth Congress alongside distinguished leaders like Peter. B. Porter, Henry Clay, and John C. Calhoun, and in 1817 was appointed senior manager of the Literature Lotteries of New York State, which required that he move to New York City. However, Chittenango remained his primary residence to which he often returned to direct the management of his extensive businesses that included flouring and saw mills, lime and plaster mills, stores, factories, and dry dock and boat yards. He also owned land near the proposed junction of the Oswego and Erie Canals, and looked upon the building of the Welland Canal as another crucial link in both countries broad chain of inland communications. It was thanks to men like Yates (fig. 4.3) that lasting economic and commercial ties were forged across the Canada–United States borderland, and that the Welland Canal was promoted and developed for the mutual benefit of both countries.

A shared interest in canals and commercial developments drew Merritt and Yates together. Born not far from one another in New York State during the latter part of the eighteenth century, both men had fathers who fought in the American Revolution. As historian Fred Landon notes, the Yates family "had fought on the revolutionary side but they were connected with the Butlers at Niagara, Loyalist of Loyalists," similar to Merritt's own Loyalist heritage, further illuminating the permeability of the border in this

Figure 4.3. **John B. Yates.** American investor John B. Yates held the largest number of private shares in the Canadian Welland Canal Company and was a keen champion of internal improvements on both sides of the Canada–United States border. Without Yates's financial and moral support, work on the Welland Canal would have been halted more than once. Courtesy of the Village of Chittenango, New York.

region.[37] During the War of 1812, Merritt and Yates served as captains in the army, though on opposite sides of the Niagara frontier. Following the war, they returned to their respective homes and plunged into farming, milling, storekeeping, and transportation projects, all the while preparing for public office in their native countries. Yates also had friends and relatives in Merritt's hometown of St. Catharines, and it may have been through this connection, or through Merritt's father-in-law, that the two men became acquainted.[38] Out of their first encounter grew a lasting friendship that was most generously manifested in the New Yorker's financial backing and personal support of the Welland Canal.[39] Though one American resented the fact that so many New Yorkers were investing American money in the Welland Canal,[40] Yates, who spent much of his life straddling the border, never let questions of nationalist sensibility or commercial rivalry get in the way of his broader regional and transnational vision.

It was to Yates's encouragement and financial support that the Welland Canal owed its existence. So important was his role in the success of the undertaking that the Upper Canadian Assembly later acknowledged that the country was "mainly indebted for the early construction of this work, he [Yates] having sustained it by his individual credit and resources on two occasions when it must otherwise have been stopped."[41] A key player in the history of the Welland Canal Company, Yates worked feverishly beside Merritt to mutually promote the Canadian project. More than once did he accompany Merritt to the capital in York to attend board meetings and other business connected to the canal. During one visit, Yates gave verbal testimony before a House committee in which he spoke of the Welland Canal's advantages. Next to the obvious benefits the Canadian people would derive from such an important work, the New York entrepreneur noted that developments on the New York canal system, particularly "the Oswego Canal, now under way," would accommodate commercial transactions on the Canadian project. He also used estimates and evidence from the Erie Canal, with which he was connected, to persuade the committee of the inherent advantages of their own great work. More than once did he express his belief that "[t]he outlet which the Welland Canal will open to the productions of the United States, as well as the Province of Upper Canada, will be highly beneficial to the interests of both countries."[42] Yates correctly prophesized that the Canadian canal would receive immediate attention in the United States, noting: "It will be profitable for American merchants along Lake Erie to use this canal, even for a New York market, and if profitable they will do it."[43] As indicated in the next chapter, it was also owing to New York's Oswego Canal that the Welland was transformed into a transnational trade artery serving both countries' commercial interests. With the backing of American investors like Yates, Merritt's dream of a canal bypassing Niagara's stupendous Falls was finally coming to fruition.[44]

St. Andrew's Day, November 30, 1824, was chosen to commemorate "turning the first sod on the Welland Canal."[45] For the past several weeks the weather had been unusually rainy, but at eleven o'clock on the celebratory morning, the rain miraculously let up and the sun came out just in time to welcome the two hundred or so well-wishers who gathered at the canal site. With his usual eloquence, Merritt expressed the conviction that the Welland Canal would be "as great a national object to the Province, as the Erie Canal to the State of New York."[46] Merritt acknowledged the progress on the Erie Canal, noting that it "will create a competition between the rival markets of New York City and Montreal—and be a general benefit to the whole country above us." Assuring his audience that there was nothing novel or intricate in the canal, Canada's father of transportation noted that New York State had proved the feasibility of a similar undertaking, and that Canada need "only follow the plan adopted by that celebrated and enlightened statesman, De Witt Clinton, and it will succeed without taxing the country one farthing."[47] Following his speech, Merritt ceremoniously unearthed the first spade full of dirt from the Welland Canal. Highly pleased with the day's transactions, Merritt and company repaired to a local inn where amid bursts of cheers and laughter they enjoyed a very good dinner.[48]

Less than two years after the ceremonial unearthing of the Welland Canal, across the Atlantic in London, Henry John Boulton, a director of the Welland Company and Solicitor General of the Upper Province, sat before a House of Commons committee to discuss the soon-to-open Welland Canal, and the neighboring Erie Canal, as rival avenues of trade. The committee asked whether there was "a tendency on the part of the western portion of the United States to deal through Montreal, supposing that our canal is finished," or is there "a tendency on the part of the Canadians to deal through the Lake Erie Canal, to New York?" Boulton honestly answered "I think there is no tendency on the part of the people to do either; I think the tendency of their minds is to send their produce to the best market, at the cheapest rate of transit." When further asked: "Which would appear to be the tendency of commerce, down the St. Lawrence, or through the Lake Erie Canal," Boulton replied: "Down the St. Lawrence in heavy articles, and in some others, because the American canal is so very long, that the expense must necessarily be considerably more in transporting articles upon it than through the Welland Canal and down Lake Ontario. . . ."[49] The solicitor made an important point that borderland merchants readily understood and championed: it was cheaper to ship heavy articles over long distances on lakes and rivers than it was canals that required hefty tolls to keep them operating.[50] Meanwhile, Boulton championed the virtues of the Welland Company in the hope of selling stock to British investors, but acknowledged the potential of alternate markets and transportation routes, no matter on which side of the international border. The solicitor's remarks that "the people in both countries are influenced by no national prejudices in sending their produce to market," and that "all the stock" in the Welland Canal "would have been readily subscribed for in New York" attracted the attention of several American newspapers including the more distant *Detroit Gazette* that championed the commercial advantages of the Canadian project.[51]

The solicitor's comment that the Welland Canal would benefit people engaged in trade on both sides of the border spoke to the interests and sentiments of canal leaders and developers in the Niagara–Great Lakes region. In New York one official found "only matters of rejoicing" in the advancement of the Welland Canal, for "the more markets which can be opened to the choice of our states the better, no matter on which side they may be of the boundary line."[52] Merritt himself noted that the question of the competing routes was "superfluous, for any person who fairly comprehends the extent of country lying on and above lake (sic) Erie, must be morally certain, that it will afford ample business for at least two channels."[53] Instead of focusing on national rivalry and conflict, Merritt was confident that the Upper Canadian project would expand and bolster commercial opportunities in the Great Lakes–Niagara region, noting that commerce, whether destined for New York City or Montreal, would find its way to the cheapest most reasonable avenue of trade. Though scholars point out that in subsequent years the Upper Canadian canal system lagged behind the New York canal system in terms of the amount of tonnage carried on each of the lines, this did not take away from the local benefits derived by borderland merchants, businessmen, and transportation leaders whose day-to-day economic and personal activities were increasingly tied to the interconnected and interlocking water system.[54] As one authority on the subject notes "there are other ways to see the Upper Canadian canal system than as a late and futile response to American enterprise and leadership." What is significant is how the Upper Canadian canal system met the local demands and needs of the people it served.[55] Recognizing the complexity of the issue, another scholar wrote "At a more intermediate spatial level, the Welland Canal carried grain to Montreal, but at the same time (thanks to the Oswego branch of the Erie) transformed both Hamilton and Toronto into termini of the Erie system, a funnel through which goods might enter but, conversely, also a funnel through which staples might leave Upper Canada."[56] As Boulton himself observed, the Welland Canal would complement the Erie Canal system particularly in the porous Niagara–Great Lakes borderland region.

An unusual degree of energy and excitement permeated the Niagara–Great Lakes borderland during the canal age. Caught up in the canal mania, William Hamilton Merritt's biographer wrote that "the stir and bustle of the great Erie Canal, then going on—the noise of whose blasting and excavating could be almost heard on our own frontier" provided an important stimulus in the building of Canada's Welland Canal.[57] Such reflections offer insight into the Erie Canal's increasing influence on Upper Canada's own transportation developments and the free flow of ideas and enterprise across the international border. This became evident in a later report of the Welland Canal Company that cited the influence of New York State's Erie Canal on Upper Canada's public works program. It was owing to this "stupendous work" that "the project now going forward, of uniting the Great Lakes of Canada by sloop navigation" around Niagara Falls was contemplated. While emphasizing the monumental importance of the Welland Canal in the commercial history of the young province, the directors disclaimed "all wish or intention of disparaging the mighty work" of their American neighbors "for to its astonishing success they are indebted for the commencement of their own" great work. The Erie Canal "excited the most earnest attention" of the Upper Canadian inhabitants, and "increased the ardor for internal improvement," for "even persons of moderate temperament were encouraged by the success which attended the opening of every mile of the New York Canal."[58]

From the start, the Welland Canal Company looked to the United States for assistance and support in their vast undertaking. Even in the makeup of the company directorship, American influence was evident.[59] A shortage of highly skilled engineers during the 1820s also caused some apprehension among the company directors. In a company report it was noted that "[a]t this period there was not a person to be obtained in Upper Canada, who knew the use of a spirit level."[60] Such a dearth of skilled professionals was not unique to Canada—the United States had faced a similar shortage of qualified engineers during the planning of the Erie Canal and, like Canada, had to turn to foreigners for assistance. Fortunately, the Erie Canal produced its own pool of native talent, which Canada increasingly drew on. Nathan Roberts was one such individual that the Canadian Company turned to. Roberts had earned the respect of North America's leading canal planners by inventing his world-famous set of double locks at Lockport, New York.[61] American engineer, David Thomas, after distinguishing himself as chief engineer on the western section of the Erie Canal, also went to Canada where he served as principal engineer on the Welland in 1826. For unknown reasons, Thomas resigned from the Welland Canal not long after, but was replaced by Alfred Barrett who, prior to coming to Canada, worked on the entire operation of the Erie Canal from 1817 to 1825. Eagerly sought after by the Welland Company, an 1826 report stressed "that no necessary expense should be spared in procuring engineers of competent ability," and so it was agreed that "Mr. Alfred Barrett, long employed on the Erie Canal" would be engaged as "the principal engineer."[62] James Geddes, chief engineer on the Erie and Champlain Canals, and Benjamin Wright, also provided invaluable service.[63] A neighboring American newspaper was gratified to state that "all the principal engineers, who assisted in constructing the Erie Canal, are eagerly sought after . . . in the neighboring province" indicating the porosity of the international border and the back and forth movement of canal engineers on both sides.[64] Even Ronald Shaw, whose groundbreaking study of the Erie Canal argued from the perspective of national particularity and rivalry with Canada wrote: "It must be admitted that nationalistic loyalties did not keep American engineers from assisting in the building of the Welland Canal."[65]

It was this type of technological and scientific ingenuity that Merritt sought when he persuaded America's key engineers to come to Canada and aid in the Welland Canal's construction.[66] With a handful of highly qualified engineers to turn to, the company next recruited contractors. The Canadian Canal Commission learned during the construction of earlier canal projects that bids could be had cheaper if tendered in the United States, so Merritt and company wisely looked south of the border for assistance on the Welland undertaking. An 1824 meeting of the Welland Canal Company recommended sending notices to several American newspapers for the letting of contracts on the Welland Canal.[67] Widespread American interest in the Canadian project was evident from the beginning. In November of 1824, a western New York paper announced that "an opportunity is presented for contracting for a job on the Welland Canal. . . . Persons who are experienced in canalling, and who are prepared with suitable tools, may make this a profitable undertaking."[68] Merritt also visited the contractors at Lockport, New York, in the hope that some of them, upon finishing the Erie Canal, might be available to work in Upper Canada. Merritt admired the progress on the Erie Canal, exclaiming: "An enterprising people can affect wonders!" Before returning home, Merritt scribbled some last minute notes in his diary:

> There is no impediment whatever in our plan; the course of this canal and all I have conversed with confirms me in this opinion; an advantage will be derived for beginning early, as many of the contractors being out of work will have all their tools on hand and prepared to commence immediately.[69]

Having heard of Merritt's progress, the *New York Spectator* sought "to congratulate our friends in that country, on the project of improvement before them. The whole scheme or project appears well designed, . . . and we sincerely wish that it may be carried on with the same spirit until completed."[70] The spirit of cross-border friendship and goodwill during the canal's heyday was evident in the *Spectator*'s praise.

Upper Canadians benefited from the American canal experience in countless ways. Originally, the Welland Canal called for "the same dimensions as the Erie Canal" but on Yates's recommendation, it was enlarged for ship navigation. In an 1824 letter to the Welland Canal Company, Yates and several of the New York stockholders urged the president to "keep in view sloop navigation . . . for unless the work contemplated is rendered truly efficient for the purposes of an extensive trade, the stockholders cannot look with confidence to realize those advantages which may be fairly anticipated."[71] A water depth of four feet was suitable for barges on the Erie Canal but, if the Canadian channel was to cater to the larger vessels navigating the Great Lakes, it needed to be expanded for ship and steam navigation, indicating the belief that both Canada and the United States had a mutual stake in the Welland enterprise. As a result of the American input, the Company enlarged the canal to a water depth of eight feet, and lock dimensions of 110-feet long and 22-feet wide. Under the revised dimensions, the Welland Canal would be capable of admitting vessels with an average cargo capacity of 90 to 120 tons, whereas boats on the original Erie Canal had an average cargo capacity of 20 to 40 tons.[72] Merritt agreed to the enlarged dimensions, commenting to his father-in-law that "by making a sloop navigation large enough to admit any vessel on Lake Erie, we will draw the transit to New York through our canal; as a vessel can sail from any point on Lake Erie to Oswego, at once."[73] Merritt's comment magnified the importance placed by the Welland Company on the New York market in bolstering the commercial success of the Upper Canadian canal system.

If Merritt's own countrymen were at first slow to embrace the Welland Canal, progress on the line increasingly drew their attention to its commercial importance. A public dinner in the spring of 1825 in St. Catharines, the home of the Welland Canal, acknowledged Merritt's "persevering enterprise and zeal towards the internal improvements of the country." The occasion must have been gratifying for Merritt who had long contemplated a canal that would connect the two great lakes and overcome "the natural barrier caused by the wonderful and well known falls of Niagara."[74] Realizing that this was no ordinary project, Merritt savored the accolades as glasses were raised and customary speeches were made in his honor. However, in addition to the customary toasts to Merritt and the Welland Canal, one other toast was deemed necessary at this happy occasion. Glasses were again raised and "the first promoter of canals in America—the Honorable De Witt Clinton!" was saluted.[75] Such complementary references to Clinton and the Erie Canal were further evidence that many Upper Canadians admired American progress and enterprise and hoped to benefit from the American example.

The actions of the Welland Company suggest that nationalist loyalties were not of paramount consideration as they came to rely more and more on the United States for help in building their canal. Of particular significance to the history of the Welland Canal was the hiring of American contractor Oliver Phelps whose earlier experience on the Erie Canal helped the Upper province in their vision of a canal around Niagara Falls. In fact, prior to coming to Upper Canada, Phelps had been summoned to Albany by De Witt Clinton who persuaded him to contract for building the Lockport Locks, along with the epic rock excavation that proved so monumental a task that Cadwallader Colden memorialized it in the "Process of Excavation, Lockport," a well-known lithograph from the Erie Canal. From Lockport, Phelps easily found his way across the river to Upper Canada following a request by William Hamilton Merritt to come to the Welland Canal.[76] The expertise that he acquired at the Mountain Ridge while contracting on the Erie Canal served him well on the Welland Canal in Canada. As noted by one scholar, Phelps's experience "at the Mountain Ridge was particularly pertinent because the Welland was traversing a route so similar in topography and geological structure."[77]

It was not unusual for canal builders to undertake jobs in both Canada and the United States, thus contributing to a cohesion and level of activity and back and forth movement across the international border. Phelps was unique among the American contractors in that he remained in Canada until his death at the good old age of seventy-two. In contrast, Phelps's youngest son Orson first found work on the Canadian project, becoming superintendent of the Welland Canal in 1830, while also assisting his father on projects like the Genesee Valley Canal in upstate New York in the latter part of the decade. But unlike his father, who desired to live his days out in Canada, Orson moved back across the river to Buffalo where he permanently settled, taking advantage of the commercial opportunities arising from the recently opened Erie Canal.[78] Noteworthy too among the American contractors who temporarily found work in Upper Canada was Judge Samuel Wilkeson of Buffalo, a leading New York State Senator and advocate of the Erie Canal who was awarded a contract for a dam on the Welland Canal while also tending to his vast projects and responsibilities in Buffalo. In fact, as early as 1825, several contracting firms employed on the Welland Canal had come from the American canal.[79] Canal builders, engineers, and their work force "moved freely between countries and regions depending on where construction was ongoing and jobs were available."[80]

The Welland Company also drew on the United States for hundreds of seasoned canal workers and mechanics who, following the Erie Canal's completion, were eager to come to Canada where employment opportunities awaited them (fig. 4.4). It helped that many of these laborers were adept in such skills as excavating, blasting, stonework, carpentry, and blacksmithing that they had acquired on the American line. Benefits were also derived from employing workers familiar with the arduous nature of canal work that involved the clearing and stripping of land, grubbing, digging ditches, shoveling through clay and rock, draining, and removing huge mounds of earth from the canal bed.[81] It is worth noting that even the oxen, cattle, and horses that hauled dirt and stamped down the soil to firm up the base

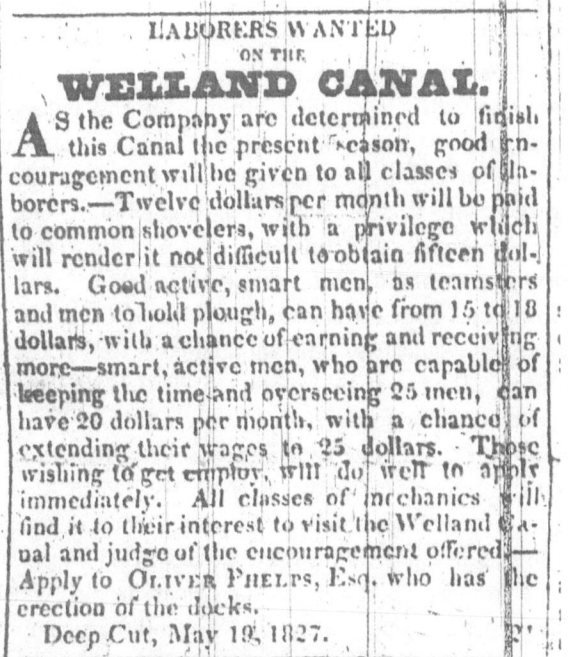

Figure 4.4. **Buffalo Journal Advertisement.** American newspapers like the *Buffalo Journal* regularly advertised employment and other related opportunities on Canada's Welland Canal. Courtesy of the *Buffalo Journal*.

of the Welland Canal were drawn in part from the neighboring United States.[82] Laborers off the Erie Canal who could bring their own oxen or horses to the Upper Canadian line were offered financial incentives by the company.[83] Apparently nationalist proclivities did not keep these men from seeking employment opportunities in Canada. The availability of American laborers, equipment, ideas, animals, and capital during the building of the Welland Canal strengthened cross-border ties and enhanced transnational economic linkages in the Niagara–Great Lakes Basin.

Canadian and American newspapers confirmed the importance of the Welland Canal in bringing economic opportunities to both sides of the border. In 1824, the *New York Commercial Advertiser* reported "upwards of one hundred houses have been erected the past year" on the Welland Canal "which are inhabited by upwards of 500 mechanics, &c, many of whom are from the United States."[84] The *Rochester Telegraph* informed its readers that the whole of the Welland line was surveyed and in readiness and that "the amount of jobs will be very considerable."[85] Local farmers and shopkeepers profited from the influx of workers and activity on the canal. An 1825 editorial in the Canadian *Niagara Gleaner* observed the rapid growth and prosperity resulting from the Welland Canal, commenting that "besides laborers on the Canal, there are blacksmiths, tailors, shoemakers, etc., there is a good market—with a constant supply of the necessaries of life, a number of merchant shops and stores with assortments equal to any in the country."[86] An Upper Canadian farmer wrote: "Our markets are well supplied; the number of hands on the canal adds to the consumption of the country and creates a readier sale than usual."[87] A subsequent letter to the St. Catharines *Farmer's Journal and Welland Canal Intelligencer* that most of the money would pass "into the United States" if Canadian "farmers and teamsters" did not apply for work, indicated the constant presence of a large American labor force on the Welland Canal.[88] As one scholar of the borderland notes, the large swarms of workers from the United States "who overran the country indicate that ambitious North American artisans were paying little attention to political allegiance when in search of a job."[89]

Since common canal laborers left little written testimony, it is difficult to get a complete portrayal of their lives on the Welland Canal.[90] However, advertisements placed by the Welland Company in both the United States and Canada offer a rare insight into the life of the workers who dug the Upper Canadian channel: "$12 per month will be paid to common shovellers (sic)," wrote one ad, and "teamsters and men to hold plough, can have from $15 to $16, with a chance of earning and receiving more."[91] Additional opportunities were offered to "smart, active men capable of keeping the time," and those willing to bring "two good yokes of oxen and a good stout cart" might earn as much as $26, indicating the ongoing difficulty of finding men and supplies like carts and equipment to work on the canal. Men were encouraged and paid higher wages if they could bring their own carts and wagons. The difficulty of finding enough laborers and the labor-intensive nature of the work is revealed in the announcement that "[a]ny person employing and bringing on fifteen good shovellers" shall be "entitled to the wages of an overseer, and hold that station" until the job is finished.[92] The lack of job security is similarly illuminated by the fact that not all workers were guaranteed employment until the job's completion. In an effort to speed up construction, contractors also devised strategies to get more work out of their laborers. Though a shoveller could expect to earn $12 per month, wages would be increased by a dollar "for every extra yard of earth they may average per day." As the primary contractor on the Welland Canal, it is likely that Phelps initiated this economic incentive—while earlier contracting at Lockport, the New York Canal Commission offered him a similar incentive in which he was paid extra if he could reduce the overall expenses on his section of the Erie Canal.[93] The Welland Canal Company even went so far as to hold contests for those who produced "the greatest number of yards" during the whole project, offering $100 for first best, second best, $90, third best, $80, and so forth.[94]

An inventory of expenses on the Welland Canal also offers a unique glimpse inside the four walls of the worker's shanty. Little more than hastily erected huts that housed the workers after hours, the Company, in an ongoing effort to economize, had the shanties manufactured from local timber supplies. Based on the inventory that Phelps submitted to the Welland Canal Company in 1827, the shanties were typically provisioned with "straw ticks" and "old blankets," "shanty tables" and "shanty benches," and basic necessities like a kettle, frying pan, cup and saucer, and utensils.[95] The rough conditions of shanty life were suggested in a letter by Merritt's wife who visited a section of the canal during the cold winter months in which, for ten miles, they proceeded through the notoriously difficult and unhealthy marsh lands "passing occasionally a few shanties, where people were at work digging."[96] Meanwhile, Phelps had brought his family from the United States to the Welland Canal where they lived "in close proximity to a long line of shanties, put up for the laborers employed on the works." The latter account was given by Phelps's daughter Charlotte many years later when she spoke nostalgically of life on the canal with her father. However, according to more than one scholar, such closeness between contractor and labor on many of the North American lines disguised a form of exploitation not unlike indentured servitude or slavery.[97] Other shelters were available for those of greater means including "several convenient boarding houses: "$1.50 per week will be paid for good common board and lodging during the progress of the work."[98] A "genteel boarding house" was also made available, suggesting the presence of women and family life on the Welland Canal.[99]

An incident in 1827 indicates that episodes of violence occurred on the Welland Canal. Fortunately for the Welland Company, the experience of American contractors like Phelps, who witnessed labor unrest on the Eire Canal, prepared them for the problems of labor discord in Upper Canada. In fact, as told by one scholar, some of the worst violence and rioting in the history of North America occurred in neighboring Lockport at the western end of the Erie Canal where Phelps earlier worked.[100] Long hours of dirty, back-breaking labor for low pay exacerbated tensions and violence on all of the lines.[101] In June of 1827, a serious fight among a handful of laborers on the Welland Canal resulted in the shooting down of one of the workers by "an officer of justice;" the laborer was not likely to live "as he was severely wounded in the face and eyes, by a full charge of musket shot."[102] Little else is known of the incident but in a bid to impose order on the line, Phelps issued a set of restrictive "Rules and Regulations" in which "quarrelling or wrangling," "profane language," and drinking and gambling were deemed major offenses. Those workers unable to forgo these habits risked being dismissed without pay.[103] The stipulation that drinking would not be tolerated by management clearly spoke to the impact of alcohol on worker violence and morale. As discussed in the next chapter, Phelps and other North American canal builders increasingly looked to the interlocking New York and Welland Canal systems to bring cross-border temperance and reform to the thousands of canal and boat men who labored on the international waterways.

America's experience in canal building impacted the Canadian system in untold ways. Even in areas of technology, America's engineers and contractors designed, or brought from the Erie Canal, various techniques, devices, and equipment to assist on the Welland. In 1827, the Directors of the Welland Canal Company "offered a reward to the person who would construct a machine that would remove the greatest quantity of earth in a given time, at the least expense." Several Canadian and American contractors competed for the prize and several innovative machines were displayed on different parts of the canal line.[104] American contractor Oliver Phelps won the substantial prize of $500 for his invention of an earth-moving machine which replaced the old system of hauling out the earth in wagons and carts, and greatly reduced the workload of men and animals. Representing some of the most advanced technology of its time, the Company applauded his ingenuity:

> The machine invented by Mr. Phelps, which has been greatly used, consists of a wheel revolving around an axle, having one end fixed to the ground, and at such an angle as to bring the rim of the wheel upon the same plane with the slope of the road up the bank; around this wheel a rope is passed, with a hook at each end, to attach the empty cart going down and the loaded one coming up . . . with little labor to the cattle drawing the latter.[105]

The advantage derived from Phelps's method was that "six teams may be attached to this machine, and work without the least inconvenience or interruption."[106] An editorial in the *Farmer's Journal and Welland Canal Intelligencer* congratulated the American contractor for his achievement:

> A large force is employed on the Deep Cut; a great quantity of earth is daily removed; and everything presents an appearance of the greatest order and regularity. The present contractor, Mr. Oliver Phelps, is certainly the most industrious and indefatigable man we ever knew . . . he has purchased and put in motion 15 or 20 machines, and about 250 yokes of cattle, besides a great number of horses, carts, wagons, etc., etc. He has also made upwards of 20 roads up the sides of the cut, for hauling out the earth; erected a large temporary building for mechanics to build and repair wagons, carts, ploughs, and other canal implements. . . .[107]

Such innovation and foresight earned Phelps "a strong claim to the confidence of the Company from the result of his exertions."[108]

The Welland Canal Company constantly benefitted from the Erie Canal experience. A tragedy on the Canadian canal in November of 1828 nearly halted the project altogether if not for American intervention and know-how. Just a few weeks before the canal's much anticipated opening, a series of slides in the Deep Cut brought the project to a complete halt. The banks of the Deep Cut literally collapsed, taking everything in its fold. A visitor to the canal in the wake of the collapse correctly blamed the "quicksands" that tended to "loosen the foundation of the canal and caused the avalanche of earth."[109] Fearing much personal and financial hardship, the Welland Canal Company called on James Geddes and Alfred Barrett, "two of the Erie Canal's most skilled engineers," who "offered a recommendation to reroute the canal, thereby removing "every apprehension of slips or similar casualties in the future" and saving the Welland Canal Company from going the way of many failed private transportation companies in the nineteenth century. The landslide at the Deep Cut received much attention in the neighboring United States. Geddes wrote to Peter Porter at Black Rock: "I have just returned from a town on the Welland Canal where I was called upon to advise respecting the disaster on the Deep Cut."[110]

American ingenuity also found its way across the border. The massive stump-pulling machine that was invented and used on the Erie Canal found its way to Canada's Welland Canal. Perhaps inefficient by today's standards, this large machine eased the burden of nineteenth-century canal work. A team of animals was attached to a rope that was coiled around a central wheel, and a chain was secured to a tree stump. When the animals pulled on the rope, the axle turned, the chain tightened, and the stump was snapped from its root and removed from the canal path.[111] The continuing need for American supplies on the canal, like ploughs, shovels, harrows, spades, picks, carts, and wagons, urged Merritt in 1825 to petition the Upper Assembly "for the remission of duties" on tools and other articles imported from the United States, as they were not available in Canada, indicating both the widespread use of American equipment on the Welland Canal and pressures to keep the border free of burdensome taxes and duties.[112] When questioned about this petition, Merritt remarked: "The proper spades and shovels are not imported here (i.e., to Upper Canada from Britain), but are made in the United States." Furthermore, "[w]agons could not be procured in the country at a reasonable price, or in sufficient quantity." Merritt's request was permitted, revealing the important role of internal improvements in liberalizing trade across the international border.[113] Even the massive amount of stone that built parts of the Welland Canal came from Split Rock, New York, not far from the Oswego Canal where the Canadian company had its own quarry.[114] In fact, it was not uncommon for laborers from both the Upper Canadian and New York canals to seek alternate employment at the quarry during down times or in winter. One young Irishman, Peter McGuire who came to Canada to work on the Welland Canal, found employment at the Split Rock quarry before seeking opportunities on the Oswego and Chenango Canals and finally settling at Syracuse, New York, on the Erie Canal.[115]

America's participation in the building of the Welland Canal also strengthened ties of commonality and friendship across the international border. American observers applauded the role played by New York's State's engineers, contractors, and workers in the neighboring province's internal improvements schemes, seeing it as a model of international cooperation and friendship between the two countries. There is one circumstance, said a New York official in 1827, that "cannot be averted to without feeling our national pride much gratified" when considering the Welland Canal:

> Both the engineer, Mr. Barrett, and the principal contractor Mr. O. Phelps, are Americans, whose professional knowledge has been gained exclusively in this state. Without laying ourselves open to the charge of boasting of our own prowess, we may fairly take the credit of having not only constructed with our own hands, the greatest work of the kind, which is perhaps anywhere to be found in the world, but for having turned out workmen as well qualified, in the opinion of those who had both countries to choose from, to grapple with similar undertakings, as the more experienced engineers of that country, to which, after all, we are free to acknowledge we are indebted for our first lessons. If, however, we can manage to improve upon her example in our national works, and also repay the original obligation by lending our citizens to assist her in works of such extensive utility as the Welland canal, we are contributing in the most direct manner to our own renown, by this tacit acknowledgement of our scientific acquaintance with matters of which a few years ago the Eastern world gave us little credit for.

"In this enlarged but perfectly practical view, concluded the above author, "we may consider the Welland Canal as another portion of the same chain which is to bind all these countries together by mutual advantage."[116] The following year, the *American Journal of Science*, which limited its annual publication of articles to twenty-eight, included William Hamilton Merritt's *Account of the Welland Canal*,[117] further illuminating the convergence of interests and friendship across the international border. The above accolades

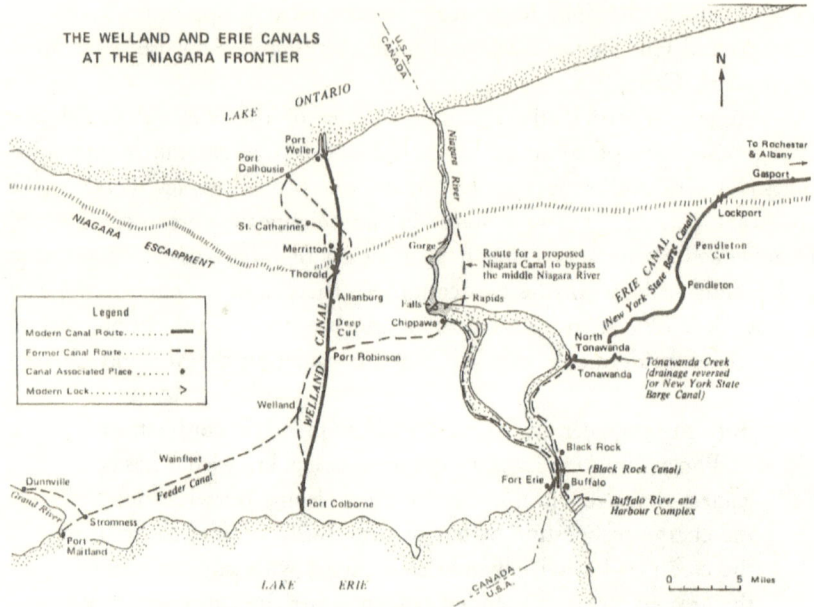

Figure 4.5. **The Welland and Erie Canals within the Niagara Borderland.** From John N. Jackson, *The Mighty Niagara: One River, Two Frontiers*.

indicate that north-south lines of transportation and communications were highly valued in the Niagara–Great Lakes region despite the emphasis on east-west connections in historical literature.

It was this type of cross-border energy and enthusiasm that drove the Welland Canal to completion. In the fall of 1829, Niles enthusiastically announced: "This great work in Canada" is said to be finished, and ceremonies to commemorate the official opening of the Welland Canal were eagerly planned.[118] Niles went on to jest that if the newly opened Welland Canal "does not much facilitate extensive smugglings, we shall be agreeably disappointed."[119] The editor of the *New York Spectator* hoped that the celebration surrounding the canal's opening "will take place in a style worthy of this important and beneficial epoch in the history of Canada."[120] With such enthusiasm and support, the canal was filled on November 30, and a British and American vessel—the *Ann and Jane* of York, U.C. and *R. H. Boughton* of Yorkstown (sic)—passed from Lake Ontario to Lake Erie. The elaborate canal procession began at Port Dalhousie along the course of the Twelve Mile Creek, and as enthusiastically detailed in the *Buffalo Republican*, after "cutting ice, in some places, three inches thick, ascending thirty-two locks at the mountain; passing the deepest of all cuts; locking down into the Welland River; sailing down that river and touching at Chippewa; stemming the strong and broad current of the Niagara River;" the vessels sailed into the "Black Rock basin through the sloop lock," where they were saluted by General Peter Porter's own steamboat the *Henry Clay*, "and cheered by the citizens." Next the vessels proceeded to Buffalo where they "were met with bursts of applause and discharges of artillery from the Terrace." As further detailed in the *Buffalo Republican*:

> To the surprise of the residents of Buffalo and Black Rock, the lake schooners *Ann & Jane* of York, Upper Canada, and *R .H. Boughton* of Yorkstown (sic), arrived in our harbor . . . having on board the enterprising projector of the Welland Canal, William Hamilton Merritt, with a company of gentlemen. The British vessel led the van. The Locks were passed on the 30th of November, ... Its progress to its termination is flattering, and the news we now communicate, that of 'the passage of vessels from lake to lake,' must be cheering indeed to the stockholders and gratifying to the inhabitants of Upper Canada.[121]

As indicated in the route followed by the *Ann and Jane* and *R. H. Boughton*, the two vessels utilized the sloop lock at Black Rock, New York, on the Niagara River during the celebratory opening of the Welland Canal (fig. 4.5). In fact, the American lock formed a functional part of the First Welland Canal until 1833 when the route was extended to Port Colborne on Lake Erie which shortened the route and terminated the use of the Welland and Niagara Rivers.[122] Prior to this improvement, William Hamilton Merritt had approached General Peter Porter about enlarging the Black Rock lock for steam vessels coming off the Welland, but the changes to the Canadian route negated this plan. Either way, the fact that the Welland Canal utilized the Black Rock sloop lock indicates that the friends of the transportation revolution in this region were not overly concerned about having a canal cut *too close to the frontier* as is traditionally argued in nation-centered history in both countries. Porter and Merritt would also later correspond about a proposal to allow American wheat and flour into Canada duty free, underscoring the transnational nature of transportation and trade in this region.[123] Following the historic passing of the *Ann and Jane* and *R. H. Boughton*, Merritt and friends repaired to the Buffalo Eagle Tavern where a few years earlier the Grand Erie Canal's ceremonial opening was also celebrated.[124] The Canadian canal directors fondly recalled during the Welland Canal's ceremonial opening that they were greeted by many of the villagers on the American side "who called to shake the hands of the navigators of the Deep Cut," and that the two vessels were welcomed at Buffalo and Black Rock "in the most friendly manner."[125]

Having at first given only light thought to "Mr. Merritt's Ditch," Canadians now recognized the momentousness of the occasion. According to Merritt, as the *Ann and Jane* made her first voyage through the canal, the banks "were crowded with people and the enthusiasm displayed on the occasion, testified that those who witnessed the display were now fully satisfied as to the prospects of the great work, which had so long occupied their attention."[126] With a glow of pride and exultation, Merritt cherished the moment as his much heralded canal was announced a success. The *St. Catharines Journal* wrote: "It cannot but be extremely gratifying to the few early and steadfast friends and supporters of this magnificent undertaking, thus to witness the complete success of their labours. . . . Falsehood and foolery cannot now prevail; and those who will open their eyes and look at the reality, can go away doubting no longer." The following season, when water was let into the feeder (originally built in 1829 to bring water from the Grand River to the canal), a celebratory dinner was "given to the contractors, laborers, etc., and a day of festivity" to acknowledge their vital role in the canal's completion.[127] In recognition of Merritt and the grand Welland Canal, the *Kingston Chronicle* recited a little ditty:

> Lo! Plann'd by one as generous as wise
> The Welland! Noble work of art we prize;
> The task undauntedly to carry through,
> Merritt was thine; be thine the honour due:
> Completed now, accept our gratified praise.
> Ontario and Erie thus made one,
> What more can hope demand, than he has done.[128]

Yet, the ceremonial passing of the two ships through the Welland Canal held only symbolic importance as work was still going forward on the Canadian line. Before long, a plan was hatched to bypass the Niagara River section of the Welland Canal so as to facilitate a more direct route between Lake Ontario and Lake Erie. The new improvement was strongly urged by James Geddes.[129] In 1833, a fleet of Canadian and American vessels heralded the completion of the new and improved line between Port Colborne on Lake Erie and Port Dalhousie on Lake Ontario, thereby bringing a formal end to the canal's construction.[130]

The Welland Canal's opening received much attention on both sides of the Canada–United States border. "The Welland Canal is at length completed," announced the *New York Albion*, "and vessels have passed from Lake Ontario to Lake Erie. We heartily congratulate the friends of that great project, and the Canadian public generally, on the final accomplishments of the scheme." The same paper asked that special recognition be attributed to William Hamilton Merritt, "whose name must always be associated with the Welland Canal, in the same degree that De Witt Clinton's is with that of the great Erie Canal." The editor continued: "We have no belief in the idea, that the Welland Canal will prove in any way injurious to the canals of New York."[131] Another American newspaper clamored in "[t]he completion of this canal will be hailed with great satisfaction by the inhabitants of New York, especially those bordering Lake Ontario, and particularly by those of Oswego, who anticipate from it, a great increase to their trade."[132]

The commercial importance of the Welland Canal to New York State was illuminated in the *New York Annual Register* in 1830. In addition to commenting on their own transportation advances, the *Register* brought attention to the neighboring Canadian channel indicating the centrality of this water system to the larger interests of New York State. "Although this Canal is located in Canada," the *Register* observed, "it has so important a bearing on the interests of this State, that it is appropriate to insert in this place the following sketch." After expounding on the canal's usefulness in accommodating both schooner and sloop navigation, thereby voiding "the necessity of transshipment from lake vessels to canal boats," the *Register* assured its readers that while the "first impression is, that this Canal will injure the western section of the New York Canal," if it is "regarded as it should be, it is not probable such a result will be produced. In consequence of the decreased price of transportation to market, arising from the facility it affords, it will induce a larger portion of the interior to use it than could otherwise afford to come to the New York market, and the increase will probably be greater than any small loss of tolls on the western section of the Canal."[133] If New York merchants and farmers felt any anxiety about the Welland Canal and the competition of the Canadian market, they did not hesitate to sell in that market "whenever it profited them to do so."[134]

American borderland merchants, businessmen, entrepreneurs, and shipping forwarders expected an avalanche of opportunities and benefits from the Canadian channel. In 1829, the *Buffalo Republican* assured its readers that the Canadian canal will "operate in favor of this place being in future a grain and produce market, as a choice of markets would be presented." In order to encourage trade through the Canadian channel, the Welland Canal Company began publishing a regular list of tolls and schedules for the American public and offered premiums to the earliest American vessels entering the canal.[135] The notion that "the Welland will rival the Erie Canal is entirely fallacious," wrote the *Buffalo Republican* observing that Buffalo's residents looked upon the neighboring canal "with complacency, certainly not alarm."[136] As one scholar observes, the Welland's complementary role in relation to the Erie Canal helps explain "the Buffalo origins of reportedly, at one moment, half the Welland's capital."[137]

Americans connected to trade along the Niagara–Great Lakes borderland profited in untold ways from the Welland Canal. When tolls became too high on the Erie Canal for example, they looked to Canada's newly opened Welland as an alternative channel for their goods. Contemplating a toll hike on the Erie Canal in 1830, the New York Canal Board warned "with the opening of the Welland Canal it will simply be a question of profit and loss; for merchandise, other things being equal, always takes the cheapest route to its place of destination."[138] John B. Yates, whose personal investments were linked to both the New York and Welland Canal systems, observed: "If the State of New York should, in order to preserve the revenue from its canal, impose an additional toll, . . . and thus attempt to force transportation through the whole of the Erie Canal, every cent of added toll would operate as a bounty in favor of the Canada trade. . . ."[139] The *Albany Evening Journal* chimed in that "the effect of these onerous tolls will be to drive large shipments into Canada. Ohio merchants, residing at Cincinnati, and its vicinity who last year received their goods upon the Erie Canal, will withdraw their business."[140] When the issue of toll increases was again raised in 1833, a correspondent of the *New York Commercial Advertiser* warned: "[F]rom what I have seen and heard, I am fully of the opinion that unless the state of New York reduces the canal toll greatly on all produce coming to the seaboard markets, the internal communications in the Canadas will be the means of drawing a very considerable portion of your western produce to Montreal."[141] Merchants, farmers and businessmen from both sides of the border learned that a toll increase could be healthy because it forced rates down on competing routes, ensuring that "everything be done on reasonable terms." All these competitions "are a benefit to the public" wrote the Canadian *Niagara Gleaner*.[142] As one authority on the subject notes, the long-term outcome of having a choice of trade routes in this region was a "gradual reduction of tolls" and a benefit to farmers and merchants on both sides of the border.[143]

The importance of the Welland Canal to local and regional cross-border trade was evident in both countries. The Welland Canal wrote the *American Banner* "which connects the waters of two lakes, Erie and Ontario, and overcomes the natural obstruction to navigation presented by the stupendous falls of Niagara, is an event of no small moment to the inhabitants of the vast regions which border upon the upper lakes."[144] Opportunities like the discovery of high quality "peat bog" on the Welland line attracted the attention of the *Buffalo Emporium*, and Buffalo merchants advertised the sale of ploughs, stoves, pot-ash, and kettles, all of which were used on the Welland Canal. Meanwhile, mill sites and town lots on the Welland Canal were advertised in the *New York Spectator*, claiming that no part of the country offered greater opportunities to American merchants, capitalists, and millers.[145] Recognizing the importance of the newly opened Canadian channel to Erie Canal traffic in the Lower Great Lakes region, a newspaper reported that a "Mr. Mecan of Niagara," has "on the way from Pendleton, on the New York Erie Canal, 48 cribs of white oak timber and staves, destined to pass through the Welland Canal."[146] In tribute to the Canadian development, the *Buffalo Republican* published a ditty called the "Welland Canal Celebration Song":

> Loud let the thund'ring cannons roar
> The *R. H. Boughton* nears our shore
> Sound the trumpets—rattle all the drums
> The *Briton*'s *Ann and Jane*, rejoicing comes . . .
> Tell the world there's MERRITT in that work,
> By which our pots and pearls, our beef and pork
> Shall find a ready sale at Montreal—
> Good gents. And Ladies fair, do please to call
> At HUBBARDS tinkering factory of Tin,
> And see what bargains there, the cash may win.[147]

The Welland Canal was a critical link in the evolving integrated, interconnecting transportation system in the Great Lakes region. In 1830, an Upper Canadian newspaper announced that "Captain Finney of the Schooner *Charles and Ann* . . . left this place . . . for Buffalo, via the Welland Canal, through the whole line of which he passed in less than 24 hours—laded his schooner at Buffalo with Pig Iron and Castings, and left Buffalo" to return to York. "Captain Finney's opinion, from the observations which he has made during this trip," is that the canal "will afford the public all the advantages which its promoters have led the country to expect."[148] Commenting on the value of new and competing channels, no matter on which side of the border, one New York paper wrote the "Advance of the West," and "the increased value of many products, and of the land," is attributed entirely "to the New York and Welland Canals." One scholar observes that as a result of the New York and Welland Canals, and the new opportunities presented along both lines, "It was a Frontier of expansion as the two canals added their considerable momentum to the existing situation."[149]

An 1831 Welland Company report expounded on the role of the Canadian canal in promoting market development in the Niagara–Great Lakes Basin and in strengthening the borderland economy:

> Trade and commerce are seeking new channels—vessels are in the course of building, adapted to its use—the demand for lumber far exceeds the supply, notwithstanding there are twelve saw mills in operation and contiguous to the line of the canal—six grist mills are built and in course of building, besides other various machinery—and a market is opened for, most bulky and useful articles . . . for the purpose of forwarding; a line of communication was kept open from Port Dalhousie to Lake Erie, by means of a steam boat which plied regularly between Port Robinson and Buffalo, and answered the double purpose of towing vessels up the Niagara River, and conveying produce from the American side. . . .

The same report celebrated the fact that trade and commerce were seeking new channels with "[l]ines of packet and freight boats" in operation "between Port Robinson and Dunnville—thus forming an almost daily communication between Buffalo, Grand River and Port Dalhousie."[150] While traveling through Buffalo, a visitor confirmed the role of transportation in generating market growth across the lakes, noting that "[t]wenty new vessels of the largest class are now building" for navigating the Welland Canal.[151] Standing on the shores of Lake Erie at Buffalo, another passerby was struck by "the show of water crafts, steamboats, the masts of vessels, and the multitudes of the canal boats" that created "the impression of a seaport, which, as sea vessels can now actually reach it through the Welland Canal, it may in some sense be considered."[152] Canada's editor of the *Niagara Gleaner* joked that American merchants placed so many commercial advertisements in newspapers in St. Catharines, the Welland Canal's home, that:

> Some of our correspondents in the northeast appear to be at a loss to know where St.Catharines is situated . . . not knowing upon which side of the [international] line it is placed. The number of advertisements by Buffalo merchants, inserted in the St. Catharine's paper, causes this doubt.[153]

Even the people of Rochester were alive to the prospects of the Canadian canal. Produce coming from the West, particularly wheat, found its way to Rochester through the Welland Canal, thereby evading tolls on the western end of the Erie Canal while also avoiding the loading and unloading of cargo at Buffalo.[154] The complementary role of the Welland Canal vis-à-vis the Erie and expanding New York canal system[155] was also recognized by one contemporary source that commented "in addition to the advantages of the Grand Erie Canal for the transportation of our products to the seaboard . . . Rochester has a direct communication by steamboats and schooners to all ports on Lake Ontario, . . . and can by this way forward its goods through the Welland Canal to the upper lakes or through the St. Lawrence to Montreal and Quebec—thus possessing a choice of markets, of essential consequences to its prosperity."[156] The north-south orientation of transportation developments, and transnational economic linkages, connected Canadians and Americans in this portion of the Niagara–Great Lakes Basin.

Economic and commercial connectivity and progress continued to characterize the canal age in this region throughout the 1830s. In 1834, *Hazard's Register* reported favorably on the Welland Canal, stating that as a result of the improved Canadian channel "the trade of the lakes has doubled within the last five years, and will continue to increase, . . . as the western . . . inland seas of 'North America' become settled, cleared, and cultivated."[157] New York investor J. B. Yates observed during the same time that numerous Americans were eagerly using the Welland and that "an important aid to the income of the canal is derived . . . by the tribute thus paid by the Americans for the use of a communication more convenient than their own. . . ."[158] The following year, the *New York Albion* predicted that an even greater part of the American trade would likely be diverted into the Welland Canal, "and Americans will doubtless pay little attention to political considerations in the attainment of commercial ends."[159] The *Buffalo*

Whig and Journal acknowledged the local benefits of the Canadian channel, noting that a timber company was being established on the American side of the Niagara River for the purpose of "carrying plank through the Welland Canal to the foot of Lake Ontario, where they are to be rafted" to the east. Wood, including staves and shingles, regularly passed through the Welland Canal from Upper Canada to Buffalo.[160] The same paper continued: "We love to see new things, new improvements, enterprises and inventions . . . every new project of our citizens or our [Canadian]neighbors, gives most ample evidence of the intrinsic[sic] value of the region around us."[161] By 1839, *The New Yorker* reported that more than 185,000 barrels of salt passed through the Welland Canal for the western market, more than traveled through the Erie Canal via Buffalo during the same year.[162] All of these accounts speak to the importance of transnational economic linkages during the canal age in the Niagara–Great Lakes borderland region.

Additional developments in inland navigation, like the opening of the Ohio Canal (1825–1832), which connected Akron, Ohio, with Lake Erie at Cleveland, brought additional commercial advantages to the Welland Canal and promoted market development and expansion on both sides of the boundary line. The opening of the Ohio Canal allowed commerce to more easily find an outlet from Lake Erie, through the Welland, and to markets in the east.[163] Niles reported in 1830 that part of the Ohio "canal is now in successful operation, and a large business is transacting upon it, in wheat, pot and pearl ashes, and other produce." This trade is "chiefly to New York, and *Canada*, via the Welland Canal."[164] In 1833, an American newspaper observed that a considerable quantity of the western produce "came in direct from Ohio, via the Welland Canal and I understand that the importation this year has increased in the article of flour, over one hundred percent, and on most articles exceeded fifty percent."[165] "Few individuals, perhaps are aware of the amount of business which is transacted upon the Welland Canal, or the advantages which we derive from the construction of that improvement" chimed in the *Cleveland Herald*.[166] A few years later the same paper assured the Welland Canal operators that "they will secure their share of the business of the lakes, as we believe every port west of Buffalo duly appreciates the importance of the Welland Canal as a channel of communication with the east."[167] In conjunction with our own canal, one Ohioan commented, "the Welland Canal will give us a choice of markets."[168] By 1839, *The New Yorker* would confirm that in one week in June "[s]eventy-eight vessels passed through" the Welland Canal, of which "twenty-five were bound for Cleveland—the latter all deeply laden with the produce of Ohio."[169] Farther west, residents from the Michigan territory looked to the Welland Canal as "another outlet for the trade of the lake country."[170] As one authority on the subject noted, Americans understood that "competition and cooperation for mutual profit were only a step apart."[171]

The strength of the Canada–United States connection was tested in the spring of 1839 when an unfortunate assault was committed against the American schooner, the *Stephen Girard* of Oswego, while passing through the Welland Canal. According to more than one account, "some intoxicated six months' men cast a volley of stones at her *fly*, and endeavoring to detain the vessel by shutting the lock-gates, considerably injuring the yawl boat slung at the stern." The captain was then ordered to haul down his flag that was shredded in the scuffle.[172] Canadian officials immediately apologized for the outrage, while several newspapers assured the American public that "everything that could be done by the Canadian authorities as an atonement, has been and will be cheerfully done to the extent of their power." As a means of making amends, the Welland Canal Company reimbursed the ship and crew including the replacement of the ship's colors. Canada's Lieut. Governour appealed to the United States, asking that the province's people not be held accountable for the reprehensible acts of "a few drunken soldiers," while earnestly promising their neighbors that all measures would be taken to assure "that nothing will again occur to interrupt the peaceable transit of American shipping through this important channel."[173] Canada's quick and humane response to the fiasco reinforced the spirit of peace and goodwill that characterized the canal age in this region. The *Stephen Girard* incident also revealed that Canadians and Americans regarded cross-border commerce and good neighborhood as paramount to their personal and economic interests.

As the next chapter will tell, the real factor in the Welland Canal's success was the Oswego Canal that connected Lake Ontario with the Erie Canal system at Syracuse, while also opening a vital international waterway between Canada and the United States. Having visited the Oswego Canal many times in the course of its construction, no one more recognized the value of internal improvements across regions and boundaries than William Hamilton Merritt who believed:

> It is a matter of little consequence to the grower in what part of the world, his produce is consumed, so long as he has to depend on a foreign market for a demand, or by what channel it reaches that market; his interest consists in the value of the

> articles at home, and any measure or any improvement which tends either to facilitate this foreign intercourse, or to lessen the expense of transportation, adds so much direct wealth to the grower, and consequently to the country.

Merritt acknowledged that open markets and transportation routes provide "a salutary and desirable competition" between the neighboring countries, a point well understood by borderland merchants, shippers, and freight forwarders in the Niagara–Great Lakes Basin. Having spent much of his life at the forefront of the transportation revolution, Canada's father of internal improvements was convinced that commerce followed "economic rather than political laws of nature."[174]

By emphasizing cooperation and connectivity, rather than conflict and rivalry, it becomes clear how the Canada–United States transportation revolution overcame natural and artificial barriers to propel transnational economic development, and as indicated in the following pages, even shape the nature of social reform and tourism in the Niagara–Great Lakes region. The Oswego Canal, in connection with the sister Welland and Erie Canals was essential in solidifying the many common linkages across the Canada–United States border.

5

The Oswego Canal

"One of the Great Traveling and Commercial Thoroughfares."[1]

The Oswego Canal, whether viewed in reference to the business done on it, or to the facilities it affords for intercommunication with the Canadas, may very justly be regarded as an important and interesting feature in the internal improvements of this state.

The above passage is from 1834, when it appeared in the *Western Guidebook and Emigrant's History*.[2] It recognizes the commercial and recreational importance of the Oswego Canal (opened in 1828) in the history of New York State and the larger North American transportation revolution. It is also representative of how North America's nineteenth-century canal leaders, merchants, freight forwarders, and travelers generally acknowledged the significance of the Oswego channel in bringing business and progress to New York State, while also facilitating closer transportation and communication links with neighboring Canada.

But this straightforward recognition faded from view in many subsequent historical accounts where the Oswego Canal's construction came to be seen as part of the age-old commercial rivalry between New York City and Montreal for the inland trade. Scholars note, for example, that after the Welland Canal's opening, and in an effort to divert trade from Upper Canada back into the Erie Canal system, New York undertook the building of the Oswego Canal that connected the Oswego port on Lake Ontario to the Erie Canal at Syracuse.³ The fact that New York State took over the building of the Oswego Canal during the same time as Canada began improvements on the Welland and St. Lawrence River system made this argument seem that much more compelling and self-evident.

While not denying the significance of the Oswego Canal to New York State or the larger nation, the Oswego's building appears far more complex when analyzed from the perspective of the Canada–United States borderland (fig. 5.1). As canal historian Charles Snyder observed, the Oswego Canal "was much more than the sum of its parts . . . it was at once a cord binding east and west" but also served "as an international waterway between the United States and Upper Canada, delivering Canadian grain and timber in exchange for American and European manufacturers."⁴ In conjunction with Canada's Welland Canal, the Oswego Canal aided New York State but also made Canadians and Americans increasingly interdependent in the postwar years. Moreover, because many canal leaders in upstate New York and Upper Canada were affiliated with the benevolent reform movement that swept through North America during the canal age, social and moral improvement also found its way through the interlocking, interconnected water system.

This chapter recounts the story of the Oswego Canals—a story far less well known compared to the many chronicles of the Erie and even the Welland Canals. In demonstrating Oswego's commercial and recreational importance in both its regional and international context, this chapter will show how this major waterway promoted economic development, market expansion, and improvement in general between the neighboring countries. More particularly, it will demonstrate the surprisingly direct role of this relatively faraway improvement in "overcoming Niagara" and advancing the regional development of the Niagara borderland, through the creation of an Oswego-Welland system linking the Erie Canal, Lake Ontario, Upper Canada and the Niagara peninsula, Lake Erie, and Buffalo.

Speaking of the Oswego Canal's advantages, preeminent North American engineer Benjamin Wright, whose reputation was made on both the New York and Canadian canal systems, remarked that "all intelligent men at an early date" looked to the Oswego "as the natural route for improvement."⁵ Scholars may have largely regarded the Oswego Canal as a mere extension or offshoot of New York State's Grand Erie Canal, but the same cannot be said of North America's nineteenth-century transportation proponents who, both before and after the Erie Canal's construction, championed the viability of the Ontario channel. In fact, it will be recalled that the heated debate over the Erie Canal's route failed to settle the question, and in fact

Figure 5.1. **A Rare View of the Oswego Canal during the 1830s.** The Oswego Canal, especially in conjunction with the Welland Canal, facilitated closer transportation and commercial ties with neighboring Canada, while also providing an indispensable link in the Northern Tour. John W. Barber and Henry Lowe, *Historical Collections of the State of New York*. Image courtesy of the Library of Congress.

magnified the commercial importance of the Oswego-Ontario channel in the public's mind. As early as 1808, renowned New York engineer and surveyor general, James Geddes, whose work proved as critical on the Erie as it did the neighboring Welland Canal, reported favorably on the Oswego channel after examining both routes for the Canal Commission. Western New York war hero Peter B. Porter believed that the Oswego–Lake Ontario channel, in conjunction with a canal around Niagara Falls, would promote commerce and recreation in the borderland region of upstate New York and Upper Canada. Even De Witt Clinton, whose political interests ultimately led him to favor the Erie Canal, acknowledged the Lake Ontario–Oswego channel as an improvement noting more than once that it was an "important link in the great chain of communications."[6] William Hamilton Merritt similarly praised developments on the Oswego which he viewed as critical to the commercial success of his own canal in Upper Canada and the development of the Great Lakes region.

Because there is so little scholarly writing on the Oswego Canal, it is worth analyzing the origins of this canal system as an important improvement in its own right before focusing on its economic, social, and recreational significance in promoting transnational linkages between the United States and Canada. Countless politicians, engineers, legislatures, and private individuals encouraged the canal's building, but it was not until the second decade of the nineteenth century that improvements to the Oswego River began. The competition over the route of the Erie Canal and the subsequent War of 1812 delayed a decision on the much publicized Oswego channel but its leading advocates never lost sight of this critical waterway. An important development in the canal's history came in 1819 with the completion of the Salina side-cut, a mile-long canal connecting the town of Salina at the upper end of the Oswego River to the Erie Canal at Syracuse. Overlapping with the opening of the middle section of the Erie Canal from Rome to Utica, the Salina side-cut cost some six thousand dollars to construct and was completed the same year. However, work on the side-cut was not without its problems. According to one report: "between July and October about one thousand men, employed on the canal, from Salina to Seneca River, were disabled from work" because of an "extensive and distressing sickness." Though the men eventually recovered, the work site "presented a most discouraging spectacle."[7] Still, in addition to boasting one of the nation's most important salt manufacturing industries that sent its profitable commerce to both American and Canadian markets, Salina, as discussed in the next chapter, would become a popular tourist attraction on the fashionable Northern Tour. Recognizing the importance of this vital commercial development for New York State and neighboring Upper Canada, Canada's *Kingston Chronicle* was one of the first papers to announce that "the Salina side cut, in connection with the recently opened middle section of the Erie Canal," would be opened to navigation in a month.[8] The New York *Ploughboy* also called attention to the development stating the "side-cut to Oswego is noticed."[9]

Even as the Erie Canal was being dug across New York State, in 1820 the New York legislature appropriated $25,000 toward a survey of the Oswego River. Two plans for the Oswego Canal were proposed—one to utilize most of the river with a short canal cut around the Oswego Falls, while the other envisioned a dug-canal paralleling the entire river. Neither was immediately adopted, but by 1823 the Salina side-cut was extended to Onondaga Lake opening a direct link between the Erie Canal and the Seneca and Oswego Rivers. Along the Niagara Frontier, Peter and Augustus Porter, who were still fighting for the Niagara Canal, were anxious to see the complementary Oswego improvement move forward.[10] Closely monitoring these developments in general, Canada's *Kingston Chronicle* also wrote that such progress "merited great praise for the enterprise and ability displayed" while further commending the internal improvements schemes as "highly honorable to the growing energies and enterprises of our neighbors in the United States."[11] As indicated in earlier chapters, and as would become even more evident with the Oswego's progress, most Upper Canadians admired the economic enterprise of the United States and welcomed developments promoting cross-border transportation and communication facilities in the Niagara–Great Lakes region. In 1823 the New York *Wayne Sentinel* brought additional attention to the Oswego Canal's progress, noting "several boats, with a large concourse of citizens passed through and entered the lake, where a salute was fired; the boats then returned to Salina, and the festivities of the day were concluded by about fifty gentlemen partaking of dinner at Beach's."[12]

At the same time, the New York Legislature incorporated the private Oswego Canal Company, marking another critical stage in the canal's history. Initially founded for the purpose of building a *hydraulic canal* only, the Oswego Canal Company had the "power to explore and designate the route of a canal, or mill feeder; and a portion of the waters of the Oswego river, may be taken and conducted out of said river, at or above the Oswego rapid, . . . so that the water can thereby be carried down, upon the east side of said river, to such point or place as may be required by those interested therein."[13] A mile or so in length, and situated on the east side of the Oswego River, the hydraulic canal was, according to one promoter "remarkably well calculated for mills of every description."[14] Continuing to ensure itself a significant role in the canal's development, the state appointed an engineer, under the supervision of the canal commission, to oversee the project. In the event that the state wished "to adopt the said canal, as part of the contemplated improvement between Lake Ontario and the Erie Canal," the act ensured that the canal commissioners had full power to make the necessary alterations. It was this clause that allowed New York State to assume eventual control over the Oswego Canal,[15] indicating not only the importance of this improvement in the public mind but the fact that the state's attention, and the nation's for that matter, was not solely on the Grand Erie Canal.

The Oswego Company directors came largely from the professional and mercantile classes. The company's first president, Alvin Bronson, virtually monopolized shipping and forwarding on Lake Ontario. One scholar observes that men like Bronson "drew upon a maritime trade ethos and international business contacts." Having arrived in Oswego in 1810, Bronson would be engaged in trade on both sides of the Canada–United States border for the next sixty years; he lived to the remarkable age of 98. Recognized for his practical experience and independent views, Bronson served in the New York Senate for several years, bringing him into contact with such men as De Witt Clinton and other known New York politicians, all of whom were connected to the nineteenth-century canal age. It was these same connections that also brought Bronson into contact with Peter Porter and company through their shared commercial interests on the Great Lakes and advocacy of the Lake Ontario route of the canal. Indeed, it was Bronson and friends who lobbied to have the canal brought down the Oswego River from Syracuse instead of continuing the Erie Canal west to Buffalo. If Bronson was by far the most recognized proponent of the Oswego Canal at Albany, back in his home town men like Peter D. Hugunin, a county judge and lawyer; William Dalloway, a town clerk; and Orlo Steele, the town's first trustee, played a role in the canal's formation. Geritt Smith (fig. 5.2), a renowned abolitionist in the United

Figure 5.2. **Gerrit Smith.** Smith was not only a leading American abolitionist and reformer, but the largest investor in the Oswego Canal and friend of open and free trade across the Canada–United States border. Octavius Brooks Frothingham, *Gerrit Smith: A Biography*. Image courtesy of Cornell University.

States and Upper Canada, and an American presidential contestant, was the canal's largest shareholder. Emerging as some of the nation's most eager canal proponents, all of these men would come to be connected to the commercial history of Oswego in the Niagara–Great Lakes borderland region.[16]

The Company shareholders were confident that the Oswego Canal would confer numerous advantages upon the village and surrounding Great Lakes community. "By the aid of this canal the navigation of the Oswego will be considerably improved. . . . We can see in anticipation, cotton and woolen manufactures, grist and saw mills, forges and trip-hammers, and various other works in operation," hailed the *Oswego Palladium*. At the same time, Oswego's residents could look forward to seeing "schooners, sloops, and boats in great numbers, loading and unloading the various products of industry, at all those factories. The busy hum of industry already sounds in our ears, and population and wealth appear before our eyes."[17] Looking ahead to the as yet completed Oswego Canal, Canada's *Kingston Chronicle* brought attention to the advantages of "the Welland and Oswego Canals over the western half of the Erie Canal."[18] As one historian observes, Upper Canadians "welcomed and encouraged developments in the United States that encouraged and facilitated communications and trade north and south."[19]

New York State would come to assume virtual control of the Oswego Canal by 1826, but the early history of the canal's construction under the private Oswego Canal Company provides a few rare glimpses into the conditions of canal life on this little-known but hugely important work. Records pertaining to the company history are scarce, but the *Oswego Palladium* offered an insight into the lives of the workers on the line in the summer of 1824. According to the paper, late one Saturday night, workmen on the canal "turned out for a spree as the phrase is, and although it did not make much of a show compared with a real Irish wake, it nevertheless disturbed the peace of society, and the prospect was that it would be continued through the Sabbath." The editor continued, however, that in fairness to the workmen they had "never saw (sic) so large a body of men conduct themselves with more propriety and good order."[20] The popular nineteenth-century image of canal workers as immoral and in need of improvement was not evident in this particular editorial describing the workers as "praiseworthy" and deserving of mention, although, as discussed in the following pages, Oswego's residents in later years would confront the prevalence of vice and alcohol on the line. Large numbers of Scottish workers from Perth, Upper Canada, similarly found employment opportunities on the Oswego Canal in 1823 and 1824, before heading back to the upper province where opportunities were soon to be had on the Canadian Rideau

Canal, again illuminating the ease in which laborers crossed back and forth over the Canada–United States border during the canal age.[21] Meanwhile, a little-known individual by the name of John Edwards of Ontario County, who until now worked as an unskilled laborer on the Erie Canal, was hired by the private Oswego Canal Company to serve as superintendent of the entire project providing a rare example of social mobility among common laborers during the North American canal age.[22] Edwards later joined the international reform movement that flowed from the interlocking New York and Welland canal systems in the 1830s.

The short history of the private Oswego Canal Company provides further evidence of the potential of canals and other internal improvements to wreak havoc on private property and the surrounding landscape. The arrival of dozens of workers at the Oswego Canal site, along with carts, wagons, carriages, tools, implements, and draft animals, invariably caused damage to the land and adjoining farms.[23] The potential of the Oswego canal to disrupt the extensive fisheries in the Oswego River was of particular concern, especially for those whose livelihood depended on that profitable business. While surveying the canal, one official noted that should the canal be built, the fisheries "will be overwhelmed by the process of damming and locking the river, and many who now draw much of their support from that source, will in consequence be deprived of their usual means."[24] Unfortunately, a dearth of records related to the private company's short history makes it difficult to determine the extent of property loss suffered or, if claims were actually made against the company, how many people were fortunate enough to be compensated. The fact that New York State held the Oswego Canal Company accountable in the first place, indicates that canal construction habitually resulted in hardships and inconveniences to the land and the people. The experience was not unique to New York. A similar misuse of people's land was reported on the neighboring Welland Canal where individual farms were laid to waste with, at least according to one report, a total "disregard of private property."[25]

As the Oswego Company made strides on its canal, developments on the Erie and Champlain Canals created a greater demand for improvements to the Oswego River. As the two canals were nearing completion, residents bordering the river and surrounding countryside petitioned the New York legislature, requesting that these two "noble works" be linked to Oswego so as to open a major commercial metropolis and an "easy, cheap and expeditious intercourse with each other."[26] The petitioners noted that such an improvement to the Oswego River "would open to our other canals a navigable coast of six hundred miles" along Lake Ontario and the St. Lawrence, "and would draw through them to our markets, much of the trade of our Canadian neighbors," again affirming the important role of transportation in facilitating commercial intercourse between the neighboring countries.[27] The New York *Gazetteer* likewise spoke to the Oswego Canal's potential in fostering trade with the provinces commenting: "The Oswego River is a valuable channel of commerce . . . vast quantities [of produce] descend this avenue to Canada." The same newspaper concluded that the "importance of the Oswego River" had been for too long overlooked, and that the policy of the state must be to improve this channel, thereby opening a fair competition, and "giving to producers and consumers their choice of markets."[28] The north-south orientation of transportation and trade was increasingly promoted by New Yorkers in this region.

With the Erie Canal nearly completed, New York State made a grand push in 1825 to extend the Oswego improvement all the way from the Oswego harbor on Lake Ontario to the Erie Canal at Syracuse. By this time, the state had taken over the project from the private Oswego Company, marking a new phase in the canal's history.[29] A legislative report, detailing the state's historical interest in the project, commented that the Oswego Canal had been "repeatedly investigated, for years past, by successive legislatures; its importance and necessity never denied; and a determination always evinced to carry it into execution, whenever a due regard to the interest of the state would permit." Accordingly, "in the opinion of the committee, the time has emphatically arrived when this great improvement should be no longer delayed."[30] At the same time that these commercial improvements were going forward in Oswego and the larger state, signs of converging economic interests with Upper Canada could be seen. Not only were New York capitalists investing heavily in the province's Welland Canal, and steamboat industry, but as expanded on later in this chapter, during the 1830s, Upper Canada's merchants would be exporting several times as much wheat across the lakes to the United States, as they shipped to Montreal. Dependency on the latter market began to decline as Upper Canada's merchants and forwarders looked increasingly to the Oswego and New York market as an outlet for their goods.[31]

On July 4, 1826, a public demonstration was held at Fulton, New York, to celebrate both the Oswego's progress and the 50th anniversary of the signing of the Declaration of Independence.[32] Despite the celebratory occasion, New York State would confront many challenges and difficulties during the Oswego Canal's building. Again, it is worth recounting some of the canal's early history as so little has been written on this vital North American waterway. Having committed to undertake the completion of the Oswego Canal, the state invested a considerable amount of time taking new levels, exploring the river and adjoining lands, and examining the projected route. According to the new surveys, the Oswego Canal would "embrace in various places, eighteen miles of the river, on the bank of which will be constructed a convenient towing path; connected with this will be a number of short canals, in all fourteen miles in length, which will form an uninterrupted navigation for boats from the outlet of the Onondaga Lake to the harbor at Oswego." Benjamin Wright, who was at the same time overseeing work on the Welland Canal Company in Upper Canada, examined the revised route and gave it his necessary stamp of approval. Governor Clinton next blessed the Oswego Canal, predicting that it would "greatly augment our revenue, and open profitable markets to our industrious and enterprising citizens within the reach of that Lake and its tributary waters."[33]

The Oswego Canal was to be divided into thirteen sections and notices of time and place for receiving bids published in neighboring newspapers. By dividing the canal into sections under different sets of contractors, the commissioners hoped to keep prices down and avoid the risks of letting the entire line to one single individual or interest. As one canal commissioner observed in 1826, "it was against public policy to have the whole of this work put into the hands of any one man or set of men, whose interests were the same, and who would form no incentives or checks upon each other in the execution of the work."[34] By letting the canal to different contractors of various skills and backgrounds, problems of dishonesty and corruption among speculators that had been evident on the Erie Canal would potentially be reduced on the Oswego. This practice was fairly standard throughout North America with the exception of Canada's Welland Canal that let out large sections of the line to one individual—American contractor Oliver Phelps—who, as recounted earlier, proved invaluable to the success of the Canadian project.

An 1826 Commission report breaks down each of the thirteen sections on the Oswego Canal, indicating the names of the contractors and the types of jobs involved. If the goal of the contractor was to finish his job or section with a profit, this was not always the case on the Oswego with its chronic financial problems and labor shortages. Competition among other works required contractors to underbid jobs, often forcing them to leave the project or declare bankruptcy. Contractors had no way of calculating the potential difficulties of a project, especially on the Oswego Canal where excessive bad weather, illness, and stoppages forced some contractors to abscond, literally deserting workers and engineers on the line. According to the Canal Commission, it was not unusual for contractors to abandon work on the Oswego Canal "owing to the low price for which it was undertaken," and in such cases, "progress has been retarded" until the job could be re-advertised and re-let.[35] The contractor's life was made more difficult by the fact that he was paid not for the entire job, but for the amount of work achieved. Typically, pricing was determined by the different kinds of jobs. On the river section of the Oswego Canal, for example, J. Canfield and S. Peck, contractors, were paid $3 a rod to build the towing path, while another team of contractors was paid 70 cents per yard to excavate rock on the Gaston Rifts in the Seneca River. More than once did contractors just up and leave the Oswego Canal when pressures became too much or they ran into unwieldy problems.[36] Peter Way notes that underbidding was usually the main cause of financial problems among contractors.[37] Similar pressures had been faced by contractors on the Erie and Welland Canals.

Contractors on the Oswego Canal were responsible for recruiting labor and caring for the needs of their work force. Workers lived at the job, and ramshackle houses were assembled along the canal path. Like the experiences of laborers on other North American lines, life on the canal was difficult, especially during periods of inclement weather and work stoppages of which the Oswego had more than its share. Demands of farming and agriculture, and competition from other North American lines, meant that local labor was often in short supply on the Oswego. A neighboring newspaper, the *Salina Sentinel*, noted, for example, that during the summer of 1827, "more than 1000 masons, and some 4000 laborers . . . will find employment" on the Rideau Canal in Canada.[38] It must have been difficult for Oswego's workers to resist offers like this or announcements like the one placed in the *Black Rock Beacon* that "high wages" on the various incomplete sections of the Welland Canal, are being offered to "shovellers, teamsters with ox teams, plough engineers, overseers of squads; and all extra work paid for, in a liberal manner."[39] The porosity of the Canada–United States border, and the constant threat of slowdowns and irregular work on the North American lines, spurred canal workers to regularly move across regions and borders in the hope of finding steady employment or more opportune jobs.[40]

An 1826 Canal report illuminates just how labor-intensive canal building was on the Oswego and the constant need for a skilled work force. In some places, the removal of trees and timber, and the excavation of a towing path was all that was required to open the river to navigation. However, on other parts of the canal, workers had to remove large rocks from the riverbed, dig through all types of soil from soft sandy loam to hard clay and gravel, and in certain areas, excavate while standing knee-deep in water. More skilled work involved the building of bridges, locks, dams, culverts, and aqueducts requiring the expertise of stonecutters, bridge-builders, quarriers, stonemasons, carpenters, and blasters.[41] No matter how equitable the contractor or job, life was not easy on the line, and in the case of the Oswego, with its constant financial difficulties and challenges, workers probably went days or even weeks without pay.

Of course, work could not go forward without the expertise of engineers who oversaw the project. The corps of engineers employed on the Oswego Canal gained most of their experience on the Erie Canal, the Welland Canal, and other North American projects. A good example of this international cadre of engineers, in addition to Benjamin Wright, was James Geddes, who first surveyed and explored the Oswego route at the state's request in 1808. Geddes's work on the Oswego overlapped with his services in Upper Canada, where he was advising and drawing surveys for the Welland Canal Company, and in Ohio where work was steadily progressing on the Ohio Canal (1825–1832). David Thomas, who served as the principal engineer on the western section of the Erie Canal, worked simultaneously on the Oswego and Welland Canals. Holmes Hutchinson also played a crucial role on the Oswego, having first been appointed engineer on the Erie Canal in 1819 and later commissioned to draw up a new map and survey of the Oswego Canal when New York took over the project. Noadiah Childs and William Jerome also served on the Oswego Canal before moving on to other canal projects. A highly mobile group, engineers crossed long geographical distances in North America to bid for new jobs, with contractors and laborers not far behind.[42]

Some of the most difficult and expensive parts of the Oswego Canal were advanced during the 1826 season, but progress was delayed by inadequate funds, financial constraints, and abandoned contracts. Still, the canal commissioners enthusiastically reported that during the season "[t]hree important dams across the Oswego River, and several guard and lift locks have been constructed; and the foundations prepared, and nearly all the materials in readiness for the completion of those to which remain to be built." From the outlet at Onondaga Lake to the Three River rift, work was completed, some ten miles in distance. Moreover, between the Three River rift and Oswego falls, a considerable amount of work had been done, and from the Falls to Oswego harbor more than half the work was completed. The whole project was in a state of considerable forwardness and the commissioners anticipated the canal's imminent completion.

Having made some noticeable progress in 1826, the builders of the Oswego Canal looked to the 1827 season with much hope and promise. If work went forward as planned, the canal would be opened to traffic by mid-summer. Anticipating a profitable season, local store merchant William Clark advertised "To contractors: 50 kegs of blasting power, and 6 bundles of Ame's Black Strapt shovels" for use on the canal.[43] However, circumstances over which nobody had control retarded the canal's completion. That summer an epidemic fever, which lasted until late autumn, engulfed most of the line. According to a report of the canal commission, the "sickness not only lessened the number of effective laborers but, by the danger and alarm which it occasioned, raised the price of labor beyond the ability of the contractors to pay, and several of them, disabled by sickness, and discouraged by the losses which they were experiencing from these unexpected embarrassments, abandoned their contracts." A superintendent in charge of building a lock, together with his laborers, were attacked by the malignant fever, and abandoned the work altogether.[44]

Numerous others on the Oswego Canal were seized by the sickness, "several died, and among them their superintendent, a most efficient and valuable man." A contractor was also taken sick and died in the latter part of October.[45] Communities adjacent to the canal were not spared; it being reported in the *Salina Sentinel* in August that the sickness had come to their village.[46] On the neighboring Welland Canal, American contractor Oliver Phelps felt so much pressure to keep his own project moving forward that he wrote a letter to the *Upper Canadian Herald* denying that the Canadian line was an unhealthy place to work.[47] So widespread was the suffering on the Oswego Canal during the fateful 1827 season that the project was almost entirely abandoned. Not until the following year did work resume on the American canal. When the *New York Daily Advertiser* announced that the Oswego Canal was nearing completion, it also reported that there was "a good deal of sickness in that tract of the country, attributed by some to the damning (sic) of the river."[48] The canals completion date seemed further out of reach, especially since the neighboring Welland Canal was luring American workers to the Canadian line with the enticing offer of "housing and free medicine."[49]

Besides the prevalence of sickness on the Oswego Canal, poor weather and storms greatly added to the difficulty of executing the work. A series of floods filled the canal, entirely interrupting its progress and adding to the expense. So profuse were the floods that "an effort was made to discharge the water from the pit with pumps, worked with horses, but the influx was found too great to be thrown out by pumps or any machinery that could be conveniently obtained and worked by horse-power. . . ."[50] Flooding became so copious that a device was invented to draw water from the canal by means of a large wheel and buckets, powered by a dam above. As detailed in the commissioner's report, the excess flooding led to the conviction "that the surest resource was to a water power which might be introduced through the canal, from a dam on the river above, and applied to the working of a large wheel, constructed with curved buckets extending to the circumference of the wheel to its center." The wheel and machinery were completed and put into operation, and work on the canal was recommenced until its final completion in the winter of 1828.

With excitement the *New York Daily Advertiser* reported that "the Oswego Canal is now completed, with the exception of a single lock, which is for the present avoided by boats by passing the place in the river. The river is used for a canal a great part of the distance."[51] Despite the daunting challenges and hardships confronted on the line, in April of 1829 the Oswego Canal celebrated its first opening season of navigation. Some 38 miles in length, the canal bordered five villages: Salina, Liverpool, Oswego Falls, East Oswego, and West Oswego. Unlike the Erie Canal that consisted of an entirely artificial channel, half of the Oswego Canal utilized the natural river, while the other half consisted of a separately dug canal. A convenient towing path was constructed alongside the east side of the canal for oxen, mules, and horses to pull the barges; and the locks and dimensions, some four-feet deep and forty-feet wide, imitated those on the Erie Canal. Structures on the Oswego Canal included twenty-two bridges, "seven culverts, one aqueduct, two waste-weirs, and eight dams built across the river; thirteen locks of stone masonry, and one of stone and timber, having an aggregate lift of one hundred and twenty-three feet, which is the difference of elevation between the marsh lands at the village of Salina, and the surface of the water on Lake Ontario." The Oswego Canal cost approximately $550,000 to build.[52] Less than a month after the canal's opening, the *Niles Register* optimistically reported that "the Oswego Canal is entirely completed, and a large business is doing on it."[53] Upper Canada's *St. Catharines Journal* heralded the occasion with the announcement that the Oswego Canal "may now be considered navigable throughout."[54]

The Oswego Canal generated much business for New York State and the larger borderland region. It augmented revenues on the Erie Canal; it promoted a hugely profitable salt industry by linking the Syracuse salt mines to the Lake Ontario outlet and Canada while also providing salt revenues to finance the New York canal system; it enhanced land values and industry along the canal's path; and it opened an extensive navigable connection between the Erie Canal at Syracuse and Oswego's bustling commercial port on Lake Ontario.[55] Indeed, even merchants and shippers affiliated with the Erie Canal soon found opportunities on the newly opened Oswego channel. In 1829, a local newspaper enthusiastically noted that "[t]he Oswego Canal has been completed for some days, and boats are now navigating it—regular lines for passengers have been established between Oswego and Syracuse." The same paper continued: "From Oswego to Kingston, Packets ply regularly across the lake, affording to travelers a cheap and expeditious mode of conveyance to Canada."[56] Profiting from the canal traffic, an Erie Canal merchant advertised that his "entire business" was being devoted to receiving and forwarding property from Syracuse to Oswego, or "any part of the United States and Canada."[57] Seeing a similar opportunity to profit from the canal system, a Syracuse merchant began to run daily lines between Albany, Syracuse, and Oswego, in addition to operating freight vessels on Lake Ontario that forwarded property to various Canadian and American ports on the St. Lawrence.[58] Meanwhile, merchants, storage and forwarding agents, and steamboat operators from Oswego aggressively advertised their services in neighboring Upper Canada. Canadian newspapers like the *Upper Canada Herald*, the *Kingston Gazette and Religious Advocate*, and the *Patriot's Farmer's Monitor* were replete with advertisements placed by American merchants hoping to profit from the Oswego's trade ties to Canada. The Oswego Canal's opening, and the beginning of a regularized steamboat service on Lake Ontario, was considered a major boon by Oswego' citizenry, especially for those with business contacts in Canada and the west.[59] Cooperation and mutual benefit, rather than national rivalry and competition, characterized the canal age in the borderland region.

However, the real boon to the Oswego Canal came with the Welland Canal's opening in 1829. The Welland Canal allowed ships to sail from Oswego to Cleveland and Detroit without breaking cargo. Not only did this benefit the Empire State, but it brought commercial and recreational prosperity to the Niagara–Great Lakes borderland economy. In fact it is not too much to say that the Oswego Canal's building and success was dependent on the Welland Canal. Having been informed in 1827 that the Canadian project was facing difficulties, the *Oswego Gazette* wrote with concern that "the grand project, from which we entertain hopes of drawing material advantages, should be delayed."[60] Oswego's mercantile community regularly read the Welland Company reports detailing the Canadian canal's progress.⁶¹ When it was evident that the Upper Canadian channel would open to trade, Oswego's *Gazette* rightly predicted that business on the lakes "will be greatly increased."⁶² A subsequent letter to the *Free Oswego* Press spoke of the expanded commercial interconnectedness of the newly fashioned Oswego-Welland line. In conjunction with the Oswego, the Welland Canal "will afford facilities for the transportation of property to Michigan, Ohio, etc., which a few years ago, would have been deemed impossible."⁶³ A few months later, a Canadian resident paid a visit to Oswego and was struck not only by the commercial excitement generated by the Oswego Canal, but by the role of the interlocking Welland Canal in bringing commercial growth and prosperity to the Great Lakes region. As seen in an item he wrote for the Canada's *Kingston Chronicle*:

> The village is situated on the southern banks of Lake Ontario—and at the mouth of the Oswego River—a rapid and powerful stream well adapted to hydraulic purposes. An extensive and substantial *Pier* has recently been constructed . . . by which means a safe and commodious harbor is formed. Several large flour mills are in constant operation and extensive forwarding store houses are erected along the wharves. The Oswego Canal which has lately been completed joins the great Erie Canal at the village of Syracuse. The completion of the Welland Canal which unites the waters of Lake Erie and Ontario will materially increase the trade of Oswego—it being supposed that a large portion of the produce coming through that channel will enter the Oswego Canal, for the New York market.⁶⁴

After detailing other improvements like a 700-foot-long bridge across the Oswego River at the head of the harbor, and new public buildings, the same visitor, writing in the above quote, pointed out that the "principal tavern is a large brick building called the 'Welland House' and is well kept by Col. White." Named after the Welland Canal, and expressing the hopes of local residents who looked to the Canadian project as an important link in their communities' prosperity and development, the "Welland House" was for a long time considered "the grand hotel of the village."⁶⁵ For Canadian transportation leader William Hamilton Merritt, the Welland House was a favorite stop on his way to New York, having many times dined at the hotel while on canal business or during visits to his wife's relatives in Utica. During one stopover, Merritt cheerfully observed that after dining at the Welland House he boarded the Oswego Canal "that was full of boats," before connecting with the Erie Canal at Syracuse in transit to Albany.[66]

August 4, 1830, was a day of jubilation for Oswego's residents. On that day, they had the satisfaction of witnessing at their port the arrival of the schooner *Erie* from Cleveland, heralding the opening of navigation on the Welland Canal between Lakes Erie and Ontario. The arrival of the *Erie* "was greeted by the ringing of bells, by a national salute of twenty four guns, by a display of all the flags in the village and harbor, and by the cheers and congratulations" of the people. The national salute was fired at precisely 12 p.m., "the American and British ensigns" were hoisted, and all the captains of vessels in the harbor were requested to raise their national flags at the same time.[67] The ceremony spoke to the spirit of friendship and good neighborhood between the United States and Canada during the age of transportation. As announced in the *Oswego Palladium*:

> Thus this great event . . . has at length taken place, and with exulting hearts, we hail the harbinger of the commerce of Erie. Another triumph of human ingenuity and wisdom is achieved. The hitherto insurmountable barrier of the Niagara is overcome, and the waters of Erie may now mingle with those of Ontario, bearing upon their bosoms the bounties of civilization, and the gifts of the arts. If there be a spot on the western waters, which, more than any other, is to reap the commercial harvest of which the Welland Canal is to be the parent, that spot is Oswego. . . . To the 600 miles of coast to which we had access, 1000 more are now added, compromising the most western counties of New York, the county of Erie, in Pennsylvania, the shores of Ohio, Michigan and Upper Canada. . . . It is needless to speculate, for imagination cannot compass the extent of that commerce which will inhabit the bosom of the northern lakes, when the regions of the west shall yield their spoils.[68]

Later that day, the celebrants gathered for a public dinner at Oswego's Welland House to pay tribute to the *Erie*. The guests celebrated the anticipated benefits of the "Oswego and Welland Canals—Nothing but state interference can disappoint the expectations of their friends." Toasts were made to "The Welland Canal—the second link in the great chain of northern commerce," and to its projector "William Hamilton Merritt—'Such MERIT can never be forgotten.'" Gerrit Smith, who held vast investments in the

Oswego Canal, toasted "The Welland Canal" as "A Stream of Wealth into the lap of Oswego."[69] Meanwhile, John B. Yates of Chittenango, New York, a patron of canals on both sides of the international border, was recognized for his "magnificent subscription to the Welland Canal," which the celebrants deemed as "proof of his patriotism and foresight." Other projects were toasted such as the "Rideau Canal" on the St. Lawrence, and the "Pennsylvania and Ohio—may the results of their enterprise, prove the wisdom of their councils." By the evening's close, the spirit of "Internal Improvements" was saluted in general: "[M]ay they prove a connecting link between us and our Canadian neighbors."[70] The *Upper Canada Herald* confirmed that "much good feeling was displayed on the occasion," illuminating the promise of transportation improvements in solidifying transnational linkages in the Niagara–Great Lakes Basin.[71] For Canada's *Colonial Advocate* the celebratory occasion was evidence of the "The first fruits" of the American canal in bringing prosperity to the borderland region.[72]

The anticipated benefits of the Oswego-Welland line did not disappoint its promoters. A month or so following the Welland House celebration, the *Oswego Palladium* announced that "the Schr. *Winnebago*, Captain Bill from Oswego, N.Y., bound for Cleveland, Ohio, laden with merchandize and salt, passed up the Welland canal, with perfect ease and safety, on Monday last. This vessel was the first to attempt passage through the Welland Canal '*with a full freight*, bound from one U.S. port to another.'" Owned by Bronson and Company of Oswego, the *Winnebago* drew 7½ feet of water, and passed through the canal with ease.[73] Having broken ground for other large lake vessels to pass, "an extensive trade, through this channel, principally in salt up, and wheat down, will be the consequence, in a very short space of time." The passage of the Oswego's *Winnebago* was also commemorated in St. Catharines, Upper Canada, the home of the Welland Canal. An editorial in the *Welland Canal Intelligencer* spoke to the commercial importance of this occasion to the province, and indeed the whole country bordering on both sides of the lakes:

> Last week we noticed the passage of the Winnabago; and now are enabled to add that of the schr. *Victory*, Capt. Hollowood, which arrived here on Monday last, from Oswego, laden with salt, consigned to a merchant of this village. This vessel belongs on Lake Erie; she passed down the Welland Canal about two weeks ago, with a full cargo, drawing about seven feet of water, continued her voyage on lake Ontario to her place of destination, and is now on her return home.

According to Captain Hollowood of the *Victory*, "several other vessels were fitting out at Oswego, when he left, to follow in the same track."[74] The Oswego Canal, in conjunction with the Welland, was promoting commerce, market development and expansion on both sides of the international border.

The hopes and wishes expressed at Oswego's Welland House in 1830 were finally coming to fruition. Almost instantly, Oswego's shipbuilding community found new fortunes in building lake vessels specifically designed for the Welland Canal. Ironically, one of the first schooners launched at Oswego in 1830 and intended for trade on the Welland Canal was called the *De Witt Clinton*. Another schooner, the *Ohio*, owned by Walton and Willet of Oswego, "was launched from the shipyard yesterday morning. She is rated at 120 tons, and is designed for the Welland Canal trade." In a series of reminiscences, Oswego resident Joshua Clark noted that Oswego's shipbuilding industry boomed thanks to the Welland Canal. Of late, he wrote, "have been launched upon the lake, numerous sloops and schooners, cutters, etc., which are doing an immense business on the lakes, through the Welland Canal."[75] The *Oswego Palladium* concurred that the navigation on the lakes and canal "is unequalled by any in the U.S."[76] The growing confidence in the Great Lakes trade was reflected in the comment of the *Oswego Palladium* in 1833: "About thirty vessels belong to this port, where some of the most considerable forwarding establishments in the state are located."[77] By 1835, eleven steamboats made regularly scheduled trips up and down Lake Ontario.[78] Two years later, the Upper Canada's *St. Catharines Journal* announced that "100 first-rate vessels carrying from 80–150 tons" have been added to the Oswego fleet and intended for the Welland Canal.[79]

New and improved vessels for use on the integrated Oswego-Welland line were proving a boon not only to the shipping and building community but to those traveling for business or pleasure across the inland waterways, rivers, and lakes. By the early 1830s, a regular packet service was connecting Oswego with Buffalo, Cleveland, and Detroit on the western lakes and Lewiston, Toronto, Kingston, and Rochester on Lake Ontario.[80] For the *Canadian Emigrant and Western District Advertiser* the Oswego Canal was considered beneficial because more emigrants were using the canal to cross into Canada from Lake Ontario.[81] In 1831, an ad in the *Kingston Chronicle* informed travelers that "The Oswego Canal Packet" will run in connection with steamboats on Lake Ontario, regularly touching at Kingston and Oswego.[82] New steamboats like Canada's luxurious *Great Britain* that made regular trips to the American side added greatly to the conveniences and advantages of lake navigation, "and must eventually be the spring of profit as well as ease of pleasure" remarked the *Oswego Palladium*. In recognition of the *Great Britain*'s first appearance in the Oswego Harbor in 1831, the spirited people of New York "gave the Captain of the *Great Britain* an invitation to a public dinner" as an expression of kindness and good will between the two countries.[83] Fast-sailing vessels like the *Rattlesnake* promised to run constantly between Oswego and Kingston "making as many trips per week as wind and weather will permit."[84] Merchant John Hamilton, a wealthy and enterprising citizen from Upper Canada, made similar arrangements "for running a daily boat between Oswego and Kingston," that would coordinate its schedule in conjunction with the arrival of the packet boats and stages from Syracuse."[85] North-south transportation and communications facilities were becoming a regular feature of the borderland.

Signs of converging economic interests were also evident as steamboats, schooners, packets, and barges crowded the lakes, rivers, canals, and harbors of the neighboring countries. By utilizing the Oswego-Welland line, merchants and forwarders could now avoid more than 150 miles in tolls on the Erie Canal between Syracuse and Buffalo, while also avoiding some of the frustrations and delays experienced on the interior route.[86] "From calculations made last summer," wrote the *Free Oswego Press*, "it is ascertained that on a ton of produce transported from Detroit to New York via the Welland Canal and Oswego, a saving of $450 would be made over the route via Buffalo and the Erie Canal."[87] During the same year, the *Oswego Free Press* wrote: "The merchants in Cleveland, and all the other ports on Lake Erie, if they are alive to their own interests, must soon observe the superior advantages this route holds out for the transportation of all their goods up from the seacoast, and their produce down to market."[88] With the exception of 1832, when cholera descended upon the interlocking canal system and virtually shut down all traffic on the Great Lakes, the steady increase in business on the canals and lakes was visible everywhere. The integrated Oswego-Welland line was proving a boon to business and expanding commercial opportunities on both sides of the border.

With enthusiasm the *Kingston British Whig* wrote about "the march of progress" in neighboring Oswego—her advantageous situation for the commerce of the lakes, and of Upper Canada being unequalled."[89] In order to better accommodate this trade, Kingston's local newspapers began regularly reporting toll rates and duties on the Oswego and Erie Canals, breaks and stoppages on the American channels, and shipping news like the *Maritime Intelligence* that was chock full of information on Great Lakes commercial traffic.[90] Canal promoters and business leaders from both sides of the border welcomed signs of an increasingly integrated and prosperous borderland economy. Oswego's *Palladium* remarked: "We are gratified to observe that the state of New York and the province of Upper Canada are beginning to discover the great mutual interests subsisting between them."[91] In jest did the *Kingston Chronicle* state, "Jonathan is wide awake when a smart carrying trade is in question."[92]

Saluting this profitable trade, the *Oswego Palladium* announced in 1835 that "our commerce will be double, if not quadruple next year to what it is this year."[93] Upper Canadian staples such as wheat, flour, pork, and timber passed through the Welland Canal in transit to Oswego and the New York market. Superior flouring mills and forwarding storehouses were erected at Oswego in anticipation of this profitable trade. "The advantages are numerous and important, and too easily understood to require comment," wrote the *Oswego Palladium*. By 1834 when the Welland Canal carried 40,634 bushels of wheat to the Montreal market, it also transported 224,285 bushels to the American market at Oswego. Over the next few years, wheat shipments from Cleveland via the Welland-Oswego channel continued to rise.[94] Meanwhile the *Oswego Palladium* continued to monitor the state of trade with Upper Canada, noting that "the arrival of produce by the Welland canal, especially wheat, is large notwithstanding the large quantities going to Canadian ports."[95] With satisfaction did the same newspaper observe that "the great mutual interests" subsisting between New York and Upper Canada were highly gratifying to those connected with the inland trade. Upper Canadian lumber was also in high demand at Oswego.[96] In 1836, the *Kingston Forwarding Company* of Upper Canada announced that they "have engaged two schooners to bring their staves from the different ports on Lake Erie, by the Welland Canal . . . these and other Schooners will take lumber from this port to Oswego, and freight from thence to the Upper Lakes." By incorporating the Oswego-Welland line, the company promised that rates would be reduced "and the public will be doubly benefited."[97]

The integrated Oswego-Welland line was proving advantageous to trade by reducing the transportation costs and expanding markets on both sides of the international border. In 1835, the *Kingston Chronicle* noted the fact that Upper Canadians could now "secure the latest spring fashions" and merchandise from New York sooner than they could through the St. Lawrence and "save a heavy investment of capital." According to the same paper, this subject has been of late "frequently discussed in private circles, and especially among our commercial friends" in the province. Neighboring New York had become "the seaport town of Upper Canada."[98] Profiting from this trade, Oswego merchant Charles Smyth advertised in an Upper

Canadian paper that he had "established a line of boats to ply between the City of New York and Oswego, expressly to secure a portion of the trade of Lake Ontario, and the River St. Lawrence, and as he hopes, of Lake Erie, by means of the Welland Canal." The same merchant added that he was "well acquainted with the New York market, and will purchase and forward to order . . . Tobacco, Coal, Tea, or any other commodity, at a very moderate commission."[99] The Oswego Canal, in conjunction with the neighboring Welland Canal, offered Canadians and Americans new opportunities as they increasingly looked to one another for merchandise and markets.[100]

Canadian and American canal leaders and visionaries believed in the practicability of the New York and Welland Canal system. In addition to improving river and lake transportation in the Niagara–Great Lakes Basin, the Oswego Canal was important in stimulating Canadian and American trade across the lakes that according to one study "tended to make the two countries more interdependent, and in the long run," close neighbors and friends. Indeed, as Canadian historian Donald Creighton observed, it was because of these commercial benefits that "the later idea of reciprocity with the United States" emerged.[101] A friend of closer trade ties with the neighboring province, the *Oswego Palladium* continued to promote open and free trade between Upper Canada and New York, believing that it "cannot be otherwise but eminently beneficial" to both countries."[102] Nor was it a coincidence that the three leading proponents of the Canada–United States reciprocity agreement during the next decade were Canada's father of internal improvements, William Hamilton Merritt; Oswego's mercantile giant and chief promoter of the Oswego Canal, Alvin Bronson; and Gerrit Smith, who was not only the largest shareholder in the Oswego Canal but also at the forefront of the humanitarian reform movement that emerged in this region during the age of transportation.[103]

While the New York and Upper Canadian canal system did much to promote cross-border commercial ties, market expansion, and development, a spirit of humanitarian reform also found its way through the interlocking canal and lake network. Much scholarly attention has been paid to the Erie Canal in spreading the reform movement's beneficent flames east and west across the canal corridor of upstate New York, but few studies consider the north-south orientation of the reform era in the Niagara–Great Lakes region. In fact many of the reform-minded individuals in both the United States and Canada were at the forefront of the canal age, believing that, in addition to improving the plight of poor canalers, cleaning up North America's canal systems and lakes would bring economic, social, and recreational benefits. Reform movements such as temperance, Sabbath-breaking, perfectionism, and even abolitionism pervaded the border. With the help of Canadian and American reformers, religious and humanitarian aid was brought to the hundreds of canal workers and boatmen who made possible the day-to-day operations of this major international, interconnected waterway. The importance of transnational and cross-border social reform and religious ferment has been obscured by the national rivalry frame that has dominated popular and academic writing on the canal age.[104]

The reform movement gained added momentum following the opening of the Oswego-Welland line. Organizations were founded to address questions of Sabbath-breaking, intemperance, and other social concerns that plagued all of the North American canals. Indeed, the problems of intemperance and Sabbath-breaking first caught the public's attention during the celebratory opening of the Grand Erie Canal in October of 1825. A Christian traveler who just happened to be in Buffalo on that memorable day acknowledged the immense importance of the work, from a commercial point of view, but worried that too little consideration was being given to its influence on "morals and religion." The canal's role "in extending the evil which is already alarming, the profanation of the Sabbath, is now visible on the whole line." While praising the efforts of several clergymen who offered "fervent prayers" during the day's festivities, the traveler worried that "if the force of religious principles" does not prevail, "the Sabbath will become a day of business from New York to Buffalo, and from Buffalo to Green Bay."[105]

As a friend of the North American canal age, Governor De Witt Clinton, had he lived long enough, would have been alarmed by the growing problems of intemperance and Sabbath-breaking on all of the inland canals. A renowned philanthropist, Clinton, in thinking about the many advantages that would come with the opening of canals and more markets, noted that farmers would no longer have to convert their grains into liquor because of the high costs of transportation, alleviating a moral dilemma in the Republic. The "opening of a market for grain will prevent its conversion into ardent spirits—the curse of morals, and the bane of domestic felicity. Whiskey now sells for eighteen cents a gallon. What a temptation to inebriety!"[106] A man whose youth was shaped by the land and religion, Clinton believed in the redemptive features of the canal age. However, as canal leaders, businessmen, and reformers learned in the years to come, internal improvements may have resulted in commercial profits for some but often came at a terrible human cost.[107]

Several organizations were founded during the 1820s and 1830s in both the United States and Upper Canada to address the social and spiritual decay of the lives of canal and boatmen on the inland waters and lakes. The American Seamen's Friend Society, founded for the purpose of improving the living and working conditions of sailors and those employed on the navigable rivers, lakes, and canals, fought against Sabbath-breaking and intemperance. Other organizations like the American Boatmen's Friend Society and the Western Seamen's Friend Society promoted reform and religious observance in New York and the western territories. Out of these organizations, Bethel operations were commenced at Oswego and other major ports in the Great Lakes region. America's Bethel and Tract societies also looked to Canada's neighboring Welland Canal as an important source of evangelic communication. As indicated previously, the Welland Canal, like the Erie and Oswego, had its share of Sabbath-breaking, alcoholism, and violence, but because of the American influence in spreading reform, the impact of social and humanitarian aid would be felt on the Canadian line.

In 1828, the *American Tract Magazine* published the accounts of a sweeping continental tour that distributed bibles to canal and boatmen on several of the North American lines. After handing out bibles to workers on the New York canal system, the missionaries proceeded to give tracts "to laborers on the Welland Canal in U.C," and to boatmen on the Lachine Canal in Lower Canada. The Seamen's Friend Society and its affiliates also engaged preachers to labor on the international canals and waterways. Tract societies "in Rochester and Lewiston and in Kingston, U.C., supplied steamboats and vessels that touched the ports with religious sustenance."[108] American builder Oliver Phelps welcomed the reform movement from south of the border. After seeing such terrible labor unrest and unhappiness among canal workers at Lockport, Phelps, in an effort to instill spirituality among his labor force in Upper Canada, asked that all workers on the Welland Canal "will feel disposed to reverence the Holy Sabbath, and as far as circumstances will admit give their attendance at the House of Worship." Phelps was also receptive to the message of nondenominationalism that characterized the religious revivals south of the border. Regarding the moral and spiritual well-being of his workers, Phelps promised that "[a]ll persons employed on the Welland Canal, let his religious tenets be as they may, shall be entitled to equal privileges and equal protection." Workers were rewarded with "[a] comfortable house . . . for the worship of God, and some good ministers of the gospel will be engaged to preach every Sabbath."[109] Toward this end, Phelps founded the First Presbyterian Church in St. Catharines, indicating management's belief that religion, alongside temperance, might improve worker welfare while at the same time hoping to promote peace on the line. A friend of religion from the United States hoped that "[t]he Welland Canal" and all "these great works, in a pecuniary point of view, will doubtless be beneficial . . . in promoting the dissemination of education, and the influence of Christianity."[110] American reformers even tried to convert the local Indian population who lived near the Welland Canal, many of them distressed by the loss of their farms caused by flooding in the canal "and by the fevers since prevailing among them."[111]

In 1830, reformers in Upper Canada and the United States united in their efforts to bring improvement to canal and lake men on both sides of the border. Toward this end, the American Bethel Society sent Reverend Gordon Winslow to spread the gospel along the Oswego, Erie, and neighboring Welland Canals. The story is told that at Oswego "on a beautiful autumnal Sabbath morn," Winslow held a Bethel service onboard the *Winnebego* and Captain Bill ceremoniously hoisted the blue and white Bethel flag before the crowd.[112] A publication of the Seamen's Friend Society reported that "[e]nergetic measures were also adopted on Lake Ontario, and the Bethel flag was hoisted at Oswego, and carried through the Welland Canal to Lake Erie." Impressed by their strides, the Oswego Bethel committee reported that "many captains and seamen, with their families, have been led to religion, and a happy and wonderful transformation seems to be going on. . . ."[113] By 1833, a preacher was regularly employed at Oswego "and the happy fruits of his labor are visible in all the ports of Lake Ontario."[114] Bethel operations also worked alongside the Boatmen's Friend Society delivering tracts and bibles and hymn books to the "friendless wanderers." Local reading rooms and libraries were opened to promote "the intellectual, social, moral, and spiritual condition" of boatmen.[115] "These decisive moments," clamored the Bethel agents, "taken in connection with the wonderful revivals that are in progress on the great thoroughfares, call loudly upon us . . . in raising up such a barrier against the flood of iniquity, which we feared they would be the instruments of bringing in upon us."[116] Apparently their work was not completely in vain, for it was reported by the Oswego Seamen Friend's Society in 1834 that their efforts were felt "not only on the thirty and forty ports of the beautiful Ontario, but down the St. Lawrence River and up through the Welland Canal, upon the shores of Lake Erie." [117]

The temperance reform movement that touched the Oswego and Erie Canals made its way to Upper Canada through the interlocking canal system. In the late 1820s, company employers throughout North America began to question the place of alcohol on the line as violence, low morale, sickness and accidents wreaked havoc on the canaler's world. The example of New York, and other American canals, where "the practice of temperance is strongly inculcated," forced the directors of the Welland Canal to confront the pernicious effects of alcohol on their own line. Citing an 1826 director's report that looked to the neighboring American canals for direction in addressing alcoholism and worker unrest, the Welland Canal Company applauded "the spirited, humane, and wise conduct of the managers in expelling from their borders the pest of tippling shops, which had infested the whole line. These nuisances, by furthering the pestiferous draught by day and by night, rendered the workmen, not only unfit for labor, but the ready instruments of riot and disorder. When drunk, they frequently fell, exposed for hours, unsheltered to the rays of the sun, and evening dews—fever and death were but too often the melancholy consequences." As a result of the American model, it became "a standing order of the Welland Canal Company to pay no bills presented for intoxicating liquors."[118]

Having witnessed some of the worst abuses of intemperance while employed on the western end of the Erie Canal, Oliver Phelps worried about the debilitating effect of alcohol on the Welland work force. In fact, it was during a trip to Buffalo, New York, where he came across an article in *The Journal of Humanity* expounding on the evils of alcohol that Phelps awakened to temperance reform.[119] Aroused by the journal, Phelps immediately rode back to St. Catharines, the home of the Wellend Canal, determined neither to imbibe himself, nor to supply alcohol to his workers. Phelps's daughter tells a story that upon returning from Buffalo, her father offered his workers an extra dollar per month if they agreed to abstain from alcohol, and those who refused were asked to collect their last pay. Apparently some of the workers left the camp following the ultimatum but returned after a few days. Then, before a gathering of his workers, Phelps symbolically poured out the contents of a whiskey barrel, and from that day forward prohibited the use of liquor on the line.[120] The centrality of alcohol in the work camps was indicated by the fact that, prior to Phelps's reform, he had erected his own distillery to help defray the costs of his worker's whiskey rations.

Closely following these developments in Canada, America's *Religious Intelligencer* was pleased to report that "[t]he use of spirituous liquors was dispensed with by about 200 men, employed on part of the route of the Welland Canal," going on to declare "[a] very manifest advance of the temperance reformation."[121] Phelps next enforced his series of "Rules and

Regulations," declaring that workers shall in no case indulge in "drinking or tippling at the groceries or taverns," especially on the day of the Sabbath.[122] The English traveler, Captain Basil Hall, visited the Welland Canal in 1827 and was struck "with the effect of the regulations which Mr. Phelps had successfully established amongst the laborers in his employment." So impressed was Hall by the imposition of "good order upon this class of laborers," that he recommended the rules and regulations be published in North America "for the use of persons engaged in similar undertakings," indicating the widespread alcohol problem on all of the North American lines.[123] In conjunction with these efforts, American engineer Alfred Barrett worked to further the cause of temperance reform in Upper Canada by helping to found the first abstinence society on the Welland Canal in 1829.[124] However, in later years, alcohol persisted on the Welland Canal despite the efforts of people like Phelps and other reformers.[125] Meanwhile, challenges continued to the New York canal system, it being reported at one point that a boatman on the Oswego Canal "died of wounds inflicted upon him while in a state of intoxication."[126] As indicated by this experience, alcohol was also an important factor in work-related accidents.

Merchants and community leaders from Upper Canada actively joined forces with their American brethren to bring improvement to the inland waterways and lakes. Having witnessed its own commercial boon since the opening of the Oswego Canal, Kingston's mercantile classes also wrestled with questions of temperance and Sabbath-breaking. The vast number of American temperance advertisements in Kingston's local newspapers during the 1830s reveals that Upper Canada's reformers were in close contact with the American movement. Allying with organizations like the Seamen's Friend Society in Oswego, Kingston's temperance leaders pledged themselves to help the hundreds of downcast men, women, and children who labored on the waterways. In 1830, the *Kingston Chronicle* wrote:

> We are indebted for the following circular from a friend in Oswego, to which we beg leave to draw the attention of our readers . . .
>
> It is proposed that a meeting of seamen and their friends, be held on the 18th of August next, in as many ports bordering on Lake Ontario and the River St. Lawrence as possible, to take into consideration the expediency of forming a Lake Ontario Seamen's Society, and if desired . . . , to appoint a delegate from each port to meet at Oswego.[127]

The society would raise funds for bringing tracts and bibles to seamen and their families living on or near the lake ports in the United States and Canada, as well as providing libraries for boats that navigated the canals, rivers, and lakes. The *Chronicle* was emphatic "that the efforts of the friends in both the British and American ports be united" to eliminate temperance and crime on the inland waterways.[128] The Seaman's Friend Society and its affiliates continued to work hard to bring reform to the many ports on Lake Ontario, the St. Lawrence, and through the Welland Canal to the shores of Lake Erie.[129]

Upper Canada's reform leaders were keenly familiar with one of the United States most prominent philanthropist and abolitionist, Gerrit Smith, whose name was regularly advertised alongside the New York Temperance Society announcements in Upper Canada's *Kingston Chronicle*.[130] Having bought up most of the land on the east side of the Oswego Canal, Smith quickly became the region's largest realty holder.[131] In addition to investing in land, Smith controlled nearly all of the shares in the hydraulic company, leased the water privileges, and erected flour mills and shops along the Oswego line. Moreover, it was as a result of Smith's spirited efforts that one of the first temperance hotels in upstate New York was founded. Actively engaged in temperance reform, Smith supported a temperance statement "to the proprietors of steam and canal boats, that they abolish the use and sale of ardent spirits on board such boats, on the waters of this state."[132] Smith may also have had ulterior motives for supporting temperance—as one of the largest commercial and realty investors in the region, he understood that profits would come from freeing the canal of vice and drunkenness. But, as a leading abolitionist, Smith was also heavily invested in transporting runaway slaves from the south who sought freedom in nearby Canada. Indeed, in later years with the passage of the Fugitive Slave Act, the Oswego Canal would carry many runaways to Lake Ontario where they crossed to Upper Canada, thanks in part to Smith's deliberate efforts.[133]

Smith was also well acquainted with William Hamilton Merritt, a close business acquaintance who shared an interest in canals, commerce, and abolitionism. In fact, many fugitive slaves who found their way to Oswego would have crossed Lake Ontario to either Kingston, Upper Canada, or to Merritt's home town of St. Catharines. Situated at the northern end of the Welland Canal, St. Catharines was a well-known terminus on the Underground Railroad. American contractor Oliver Phelps, who as earlier discussed promoted reform on the Welland Canal, also joined Smith and Merritt in the abolitionist cause. True to their mission, Merritt and Phelps are credited with encouraging black settlement in St. Catharines by offering land on reasonable terms.¹³⁴ As is well known, St. Catharines also became the permanent home of famed abolitionist Harriet Tubman who guided numerous slaves to their freedom in Canada. The transnational nature of the reform spirit that found its way through the interlocking New York and Upper Canadian canal system offers an important example of the beneficent and friendly intercourse between the neighboring countries, and further evidence of the north-south orientation of reform in this region.¹³⁵

One of the more novel cross-border reform efforts was to turn the interlocking canal and lake corridor into a sort of traveling school of science and education. In September of 1830, the *Kingston Patriot* noted that a curious canal packet towed by the *Martha Ogden* from Oswego to the Kingston harbor was met by a great deal of cheering and attention among Canada's residents. "We have heard and seen much of the enterprise of the Americans," wrote the Kingston *Upper Canada Patriot*, "who are a people resolutely bent on living as long as they can. Who but a Yankee would have thought of a floating museum?" The boat was fitted up with "a cabinet of curiosities of specimens of Geology, Orinthology, and Entomology . . . to be viewed through a magnifying medium, which renders them very interesting."¹³⁶ The idea of a traveling school of science and observation originated in 1826 when students from the New York Rensselaer School first went on a lecture series aboard an Erie Canal packet from Troy to Niagara. During one of their stops, the traveling scholars visited a school named "Polytechny," later changed to Yates Polytechnic Institute after its founder, John B. Yates of Chittenango, New York, who, besides opening one of the first schools in the United States for the instruction of the laboring classes, was a keen supporter of the Erie and Oswego Canals, as well as the largest investor in Canada's Welland Canal.¹³⁷ Gaining a great deal of attention wherever it went, the floating school of science promoted education and learning by traveling the New York and Upper Canadian canal systems and lakes. Commenting on the importance of the floating school to the dissipation of knowledge and learning in the Great Lakes region, the *Kingston Patriot* hoped that "such laudable exertions never want encouragement."¹³⁸ The common experiences of the canal age helped promote education and reform on both sides of the border.¹³⁹ To this day, the interlocking Canada–United States canal system continues to promote education and the natural sciences as a vehicle of learning.¹⁴⁰

In 1837, the *New York Farmer* recognized the Oswego Canal as "One of the great commercial and traveling thoroughfares" in North America.¹⁴¹ If the Oswego Canal, especially in conjunction with the sister Erie and

Welland Canals, helped stimulate commerce, expansion, and progress in general, another beneficial offshoot of the interlocking canal system was the creation of a thriving tourist industry in the Great Lakes–Niagara region. As indicated in the next chapter, during the second quarter of the nineteenth century, numerous travelers embarked on what was familiarly known as the Northern Tour, a leisurely excursion through the northeastern United States and Canada with Niagara Falls being the most popular destination. Cheaper and more convenient travel options ushered in by the canal era, and safer and more reliable transportation that came with the cleaning up of North America's waterways, helped stimulate the tourist boon inland. However, in addition to these improvements, it was also a new-found interest in canals and other internal improvements *in their own right* that drew so many people to the region. Indeed, for the curious spectator, Niagara Falls offered not only a glimpse of one of the world's greatest wonders, but some of the most important developments in canal technology and innovation. Alongside economic and commercial development and progress, even the nature of tourism and improvement in general benefited from the converging American-provincial transportation network.

6

"The Great Wonder of the Canal"

The *Great Britain*, on her downward trip yesterday, had among her passengers a great number of American ladies and gentlemen, travelling the northern tour.

—*Upper Canada Herald* (July 28, 1835)

The creation of a thriving cross-border tourist industry in the Niagara–Great Lakes region was an additional benefit of the North American canal age. Having recognized the potential of canals and other internal improvements to encourage cross-border commerce, openness, and improvement generally, Canadian and American hotel operators, businessmen, merchants, and entrepreneurs turned their attention to the tourist trade long centered at Niagara. The story of this attraction, and tourism generally, has been seen through the lens of the natural wonder of the Falls itself, and the natural border it marks between the United States and Canada. Just as nation-centered history has shaped much writing on the canal age, scholars have discussed the tourist industry as distinctive and separate United States and Canadian stories. National borders have largely defined studies of nineteenth-century tourism and transportation between the two countries, despite the commercial, social, and recreational complementarities of the canal age in this region.[1] Scholars recognize for example the importance of the Erie Canal in transforming Niagara Falls into a fashionable tourist resort, but have been much slower to notice how the integrated New York and Welland Canal systems promoted a cross-border tourist experience on a grand scale.

This chapter will show how all three canals—the Erie, Oswego, and Welland—promoted and shaped a cross-border tourist boon at Niagara. Canals, locks, and other internal improvements helped overcome natural and artificial barriers in the process transforming the Niagara–Great Lakes region into North America's premier tourist destination. Indeed, the sources suggest how these man-made wonders were as fascinating to visitors as were natural wonders like Niagara Falls, and a big part of the visitor experience.[2] When viewed from the vantage of the borderland, and transnational stories explored in previous chapters, tourism and travel appear as major expressions and facilitators of greater Canadian-American cooperation and communication—bringing the themes we have been tracing through the overall canal era into sharp and comprehensive focus.

Improvements in transportation transformed the face of tourism in upstate New York and Upper Canada, strengthening the borderland economy and cross-border relations generally. Much of this tourist boom took form in the popular Northern Tour, a leisurely trip across the northern United States and Canada that culminated at Niagara Falls. New vehicles on land

Figure 6.1. **Dwight's Map of the Erie and Welland Canal System along the Niagara Frontier.** As indicated by Dwight's map, man-made wonders like the Erie and Welland Canals were central to the visitor's experience at Niagara during the nineteenth-century age of transportation and innovation. Theodore Dwight, *The Northern Traveller*. Image courtesy of the Library of Congress.

and water, reduced fares, and greater ease of travel and comfort fed a new fascination with canals and other internal improvements in their own right—all of these encouraged a flood of traffic to Niagara Falls and the surrounding Great Lakes district.

To reach Niagara, the northern tourist followed one of several routes inland. For the large number of English and European travelers who sailed to America as part of the tour, the principal port of entry was Sandy Hook where, in 1825, Governor De Witt Clinton ceremoniously poured a keg of Lake Erie water into the Atlantic commemorating the Grand Erie Canal's official opening. Travelers made their way to Albany (at the far eastern end of the Erie Canal). Advances in steamboat navigation on the Hudson River after 1815 meant that the entire journey from New York City to Albany could be made in approximately ten hours, including several stops along the way. Included in the fare were visits to famous attractions like West Point, the site of the United States military academy with monuments to fallen soldiers like Colonel Eleazer Wood, a distinguished officer who died in Fort Erie, Upper Canada, during the War of 1812. That the final leg of their journey allowed travelers to visit the very site at Fort Erie where Wood fell must certainly have heightened the excitement for visitors.[3] Other points of interest included Hyde Park, the home of Dr. David Hosack, a well-known New Yorker who just happened to pen the first biography of De Witt Clinton.[4] After arriving at Albany, travelers might follow the direct westward path of the Erie Canal all the way to Niagara, or opt to go north to Lake Champlain, crossing to Montreal on the St. Lawrence, then heading west to Kingston and Niagara, before trekking back by way of Lake Ontario and Oswego.[5] Transportation choices during the canal era made the tour less time-consuming and that much more accessible and interesting. In the twenty-first century, travel companies have reinvented the idea of the Northern Tour, renewing the once historic and recreational attraction of these interlocking canals and waterways.

Though much attention has been paid to the Erie Canal in both revolutionizing travel and popularizing the Northern Tour, part of the pleasure of canal age travel was combining the various modes of land and water transportation in transit to Niagara Falls. In fact, from Albany to Schenectady, on the first section of the Erie Canal, stagecoaches often proved more convenient because of the excessive number of locks and stoppages on this part of the line. Coaches were so popular on this route that as many as thirty a day made the round trip.[6] During winter travel, the stagecoach was conveniently placed on a sledge, and converted into a sleigh which according to one passenger made travel more pleasant and expedient.[7] By 1831, the introduction of rail on this section of the canal saved a day's travel-time and offered the thrill of a new and promising mode of transportation. In fact, Canadian canal promoter William Hamilton Merritt was one of the first to experience the historic rail between Albany and Schenectady, recalling that "the rails were of wood, covered with a band of iron. The cars were drawn by horses, on the plateau between."[8] Excited by this innovation, Merritt returned to Canada where in time he spearheaded, in cooperation with the United States, a pedestrian and commercial railroad spanning the Niagara River.[9]

In 1824, the *Rochester Telegraph* noted that at no former time had travel been so opportune: "The facilities which are offered for traveling through the country, in every direction, and the reduced fare on the public stages and steam-boats, induces many to embrace the opportunity of leaving the cities . . ." for a jaunt through the countryside.[10] In transit to Niagara Falls in 1826, one northern traveler observed that the route "can be diversified as often as a person chooses."[11] Others could travel "from Schenectady to Utica by canal, and from Utica to Rochester by stage, and then take the canal again to Lockport or vice versa, changing the mode of conveyance at the end of each stage."[12] It was even possible to abandon the Erie Canal and stagecoach altogether as indicated by the experience of one traveler who opted to ride a saddle horse all the way to Geneva before purchasing a horse and wagon and proceeding to Lockport.[13] Another extraordinary account told of a man who crossed the entire length of the Erie Canal between Schenectady and Buffalo, *on skates*![14] Whatever the mode of conveyance, nineteenth-century travel guides such as the *American Traveller* helped plan the Northern Tour, offering useful information on coach, packet and steamboat schedules, fares, and other travel arrangements on both sides of the border.[15] On the 110th anniversary of the War of 1812's conclusion, a Canadian sportsman opted for a more novel mode of transportation by swimming the entire length of the Erie Canal in recognition of the historic occasion![16]

Daredevil's aside, for the avid canal enthusiast, the 363-mile-long Erie Canal with its renowned technical and engineering innovations was itself an attraction not to be missed. Even before its completion, travelers came from afar just to marvel at the celebrated locks and bridges and aqueducts along the canal's route.[17] In transit to Niagara in 1821, Catharine Sedgwick stopped at Utica to observe the canal, believing it the most stupendous monument "of the enterprise, industry, resolution, and art of man." Further west, Sedgwick rode on a recently opened section of the canal, commenting, "[t]he boat is drawn by two fine horses; the hindermost has a rider. They go on a very fast walk, at the rate of four miles an hour, including stops, which recur every eight miles. The canal is forty feet wide and four feet deep." Sedgwick's journal indicates a fascination with canal technology and labor in general. "We saw all along the opposite bank of the river a multitude of men, at work on the canal," she wrote. "We are told more than one thousand between this place and Albany." Near Utica, she watched a boat pass through the lock of "the little canal which was made here more than twenty years since." Sedgwick was referring to the Little Falls Canal constructed by the Western Company some thirty years before the Erie Canal's opening. As she observed this historic work, Sedgwick recalled with gratitude "the patience and interest" with which the lockkeeper "explained to me the construction and operation of the lock." Celebrating the Erie Canal's progress, the New York canal commissioners took a leisurely ride on the recently opened middle section, commenting that "the novelty of seeing large boats drawn by horses, upon waters artificially conducted, through cultivated fields, forests, swamps, over ravines, creeks and morasses, and from one elevation to another, by means of ample, beautiful and substantial locks, has been eminently exhilarating."[18] The novelty of watching "the mechanical rising and falling" of locks also had its pleasures. Caroline Gilman, who usually complained about the discomforts of canal travel, did concede that "passing the locks, too, has its attraction. . . ."[19] These accounts confirm that canals were not just about experiencing the aesthetic or romantic, but had become interesting attractions in themselves.

Canal travel was more than just anticipating the scenic wonders of nature or the sublime (see fig. 6.1). At Syracuse, for example, Englishman Basil Hall took pleasure in observing the canal that was "thickly covered with freight boats and packets," and "was amused by seeing amongst the throng of loaded boats, a gaily-painted vessel lying in state, with the words *Cleopatra's Barge* painted in large characters on her broadside."[20] While traveling through New York in 1829, Canadian merchant and journalist, William Lyon Mackenzie, who regularly extolled the virtues of De Witt Clinton and the Erie Canal, passed the time by recording the conspicuous names of canal packets like the *General Jackson*, the *Lafayette*, the *John Hancock*, the *Napoleon*, and his own boat, the *Buffalo*. Although such epitaphs amused passengers, by law New York State required captains to print the name of their boat "in letters at least four inches long" for purposes of collecting tolls and registration.[21] Meanwhile, Mackenzie was struck by the flurry of activity on the canal, writing: "Our boat has already passed twelve or fifteen other larger vessels crowded with passengers. . . ." Frequent stops allowed passengers to get on and off the canal and contributed to the excitement of travel. While some passengers remained on the canal for the entire trip, others stayed on "only for a mile or two, paying in proportion to the distance."[22] James Stuart observed with interest that besides passing travelers, workmen "were often with us for only a mile or two" on the canal.[23] For Nathaniel Hawthorne, who rode the Erie Canal in 1835, there was as much entertainment in the variety of boats and the dexterity of the boatmen as there was in the bordering landscape:

> Sometimes we met a black and rusty-looking vessel, laden with lumber, salt from Syracuse, or Genesee flour, and shaped at both ends like a square-toed boot, as if it had two sterns, and were fated always to advance backwards. . . . The most frequent species of craft were the 'line boats,' which had a cabin at each end, and a great bulk of barrels, bales, and boxes in the midst; or light packets, like our own, decked all over, with a row of curtained windows from stern to stern. . . . Once we encountered a boat, of rude construction, painted all in gloomy black and manned by three Indians. . . .[24]

Though often overshadowed by the Erie Canal in the history of transportation and tourism, the newly completed Oswego Canal was a popular choice of travel on the Northern Tour, and contributed markedly to the trans-border nature of the tourist industry in the Niagara–Great Lakes region.[25] A cursory glance at scholarship, however, indicates that the Oswego has been entirely passed over in favor of the Grand Erie Canal as a major cultural and recreational phenomenon. Just as the broader commercial and international significance of the Oswego Canal has been understates in the history of American transportation, so too has its recreational role in promoting and selling the Northern Tour. In fact, for the hundreds of tourists who embarked on the inland journey, it was possible to avoid the entire western section of the Erie Canal by taking the recently opened Oswego route which connected to Lake Ontario and Niagara Falls. A letter in the *Oswego Press* prophesized, for example, that the newly opened Oswego Canal would soon become "an indispensable part of the 'northern tour.'"[26] Another enthusiastic traveler wrote, the most expeditious route from Albany is "to proceed to Oswego from Syracuse, by a packet boat on the canal; there they will find good hotels—of the Welland House I can speak."²⁷ As previously discussed, Oswego's aptly named Welland House gave further recognition to the transnational linkages in this region, and the role of the Welland Canal in bringing commercial, beneficent, and recreational progress to Oswego's thriving village. The *British Whig* believed that more travelers to Niagara Falls "who now go by stage or the Erie Canal, would pass by the beautiful village of Oswego," if they were more aware of this recently completed route.²⁸

Within a few years of its opening, the Oswego Canal had become an indispensable part of the Northern Tour. By 1835, travelers were strongly encouraged to leave the Erie Canal at Utica or Syracuse and board an Oswego packet or steamboat for Lake Ontario. The *Oswego Palladium* attested to the canal's growing popularity, noting "the amount of travel which has passed through this village the past season . . . is all together without precedent." So popular had the Oswego Canal become that one fellow tourist was disappointed that he did not know of it as it was now regarded as "the best route to the falls."²⁹ The Oswego Canal was fast becoming the preferred route of travel to Niagara Falls and Canada.

The Oswego Canal offered many novel attractions. In particular, tourists anticipated seeing the Salina salt works, the commercial growth of which was attributed to the Oswego Canal and a growing demand for this

View of a field of salt-vats, Salina.

Figure 6.2. **Salina Salt Works.** During the nineteenth century, Salina became one of the nation's most important salt manufacturing industries to send its profitable commerce to both American and Canadian markets. Revenue from these sales helped finance the New York canal system. Salina was also a popular stopover on the fashionable Northern Tour. John W. Barber and Henry Howe, *Historical Collections of the State of New York*. Image courtesy of the Library of Congress.

precious commodity in Canada and the remote regions of the west. Situated at the northern end of the Oswego Canal, the salt works were becoming an indispensable stop on the Northern Tour. While traveling the Oswego Canal in 1832, the Rev. Isaac Fidler was given a lecture by one of the salt proprietors "who explained the various particulars of his business" to me.[30] By the mid-1830s, the salt works drew a greater concourse of people to the region. "From Syracuse to Oswego," wrote *Atkinson's* "there is a direct line of comfortable canal packets, which run the distance in about eight hours. The canal passes through the extensive salt works at Salina, which may be readily examined while the boat is passing through the locks. Oswego is well deserving of a visit, as it occupies a very important position on the lake, . . . we believe this line to be cheaper, more expeditious, and decidedly more agreeable than any route to the Falls now traveled."[31] Captain Frederick Marryat, a famous English Navy officer and novelist who included Charles Dickens among his closest acquaintances, boarded a packet on the Oswego Canal at Syracuse in 1837, observing "Salina is a village built upon a salt spring . . . the banks of the canal, for three miles, are lined with buildings for the boiling down of the salt water, which is supplied by a double row of wooden pipes. Boats are constantly employed up and down the canal, transporting wood for the supply of the furnaces" Marryat was especially struck by the international nature of the salt trade noting: "Two million bushels of salt are boiled down every year . . . and transported by the canals and lakes to Canada, Michigan, Chicago and the far west."[32] Tourist manuals also encouraged stops at the salt mines where the elaborate manufacturing could be witnessed:

> Salina is a mile and a half from Syracuse. The first salt-spring was discovered at this place by the Indians, from the circumstance of its being visited by deer and other animals. The *Oswego Canal* commences at this place, the surplus water of which is used for the purpose of forcing (by a powerful hydraulion) the salt water 85 feet up the hill into a large reservoir. It is forced up at the rate of 300 gallons per minute, whence it is conveyed by logs to the factories in the neighborhood, which amount to 175 within a circuit of seven miles. The springs and works all belong to the State, to which the manufacturers pay imposts of 30 cents per barrel of 5 bushels, which are applied, according to the constitution of the state, towards discharging the canal debt.[33]

Such accounts illuminate the role of salt duties in financing the New York Canal system, and are a reminder that the Oswego Canal, with its row after row of manufacturing salt works (see fig. 6.2) dotting the landscape, was fast becoming one of the most popular attractions on the Northern Tour.[34] A New York State tourist who stopped at the Salina salt works noted: "Many strangers here prefer to leave the Erie Canal and go to Oswego, and thence by steamboat to Lake Ontario to Niagara, by way of variety."[35]

The Oswego Canal was increasingly heralded as "the most expeditious route to Niagara Falls and Upper Canada."[36] In addition to the novelty of boarding an Oswego packet, a trip along the canal offered an opportunity to engage in some leisurely sport like fishing: "those who are fond of fishing will find capital fish in the Onondaga River," the *Atkinson's* press assured its readers.[37] Boat captains even entertained passengers by spearing salmon near Oneida and Onondaga Lakes, in route to Oswego and Lake Ontario.[38] With assurance did *The New Yorker* claim that the Oswego Canal "is perhaps preferable at present to the inland one" and travelers "will do well to take it."[39] "The great increase of traffic through this village" announced the *Oswego Herald*, "shows conclusively that the public are becoming acquainted with this shortest, most expeditious and cheapest route" to Niagara Falls, "saving at least one day in time, and $3.50 in fare by this route."

Having once reached the Lake Ontario terminus at the foot of the Oswego Canal, tourists were drawn by the bustle of the Oswego Harbor constantly under improvement. A stroll across the expansive 700-foot-long wooden bridge, connecting the two banks of the Oswego River, or a stop at the Oswego lighthouse with its keeper's dwelling that remarkably still exists today, were now must-see stops on the northern itinerary. By 1838, the Oswego Canal with all its offerings was receiving "a great portion of the fashionable northern and western travel."[40] At the Oswego port on Lake Ontario, travelers also got to enjoy the commotion of international packets, steamboats, and schooners, many of them waiting to take passengers on the next leg of their journey to Niagara Falls and Canada.

Travelers were eager to experience the new and improved steamboats that regularly plied back and forth between the United States and Canada. Conceived for both commerce and pleasure, steamboats carried passengers, immigrants, merchandise, and mail while pulling barges, rafts, and other sailing vessels on the inland waters. Competition among canal and steamboat operators on the Great Lakes meant greater conveniences and improved services for the northern traveler. In 1830, the proprietors of the steamboat *Ontario* assured their passengers that "No accidents of a serious nature, have ever happened to steamboats propelled by two low pressure engines" of which the *Ontario* boasted. The operators of the *Martha Ogden* guaranteed regular trips between "Oswego and Niagara, Prescott, Kingston and Brockville," assuring their patrons that "many improvements have been made to the steamboats which will add to their safety and expedition, and to the comfort of passengers."[41] The elegant Canadian-owned *Great Britain* offered first-class passenger service between Oswego and Niagara Falls. The trip "will be performed in between 12 to 14 hours, which will be a saving of nearly half the time when compared with the old route of the Erie Canal, to say nothing of the superior accommodations and pleasantness of the route, as well as the saving in expense."[42] The New York State Tourist guide wrote that many travelers preferred the Oswego Canal over the Erie Canal because of its "varied and lively scene": Steamboats ply to the various ports on the lake between . . . Kingston, Sacketts Harbor, Oswego, Toronto in Canada, the mouth of the Genesee near Rochester, Queenston and Lewiston, and offering every facility to numerous travelers and strangers to vary their route either by railroads, canals, and steamboats.[43]

For the northern traveler, steamboats on Lake Ontario were not only "very exciting,"[44] but allowed travelers to avoid the long western section of the Erie Canal. For some, the prospect of remaining on board the Erie Canal "for twenty two hours was not very agreeable."[45] Steamboats meant greater comfort, speed, and reliability, while also helping to vary the long journey inland. American and Canadian steamboat operators mutually promoted steamboat travel on the Great Lakes. Steamboat operators frequently advertised races between the competing lines. In a spirit of friendship and good neighborhood, the *Kingston Chronicle* spoke in 1837 of a "Mammoth Race" between the splendid steamboats *Great Britain* and the *United States*. Steamboat racing was an amusing tourist diversion, but because of dreadful accidents, passengers liked to be told beforehand whether captains intended to partake in trials of speed between competing operators.[46] At 380 tons burthen, 145-feet long, and two 60-horse power engines, the American-owned *United States* was said to be so fast that she "obtained the familiar title of 'the Clipper of the Lakes.'" She was also quite luxurious with staterooms and berths and a promenade deck that spanned the entire length of the ship. However, during the much-anticipated race, the *United States* did not live up to her stellar reputation. In the wake of the *Great Britain*'s victory, the *Chronicle* heartily cheered "Hurrah, Hurrah" for the *Great Britain*—"We guess Mr. Jonathan will hardly challenge Mr. Bull in a hurry!" Ironically, as it turned out, the captain of the victorious *Great Britain*, a highly skilled seaman by the name of Joseph Whitney, was an American who not long after returned to the United States to command the *United States*! The enduring sense of cross-border community and friendship between the neighboring countries was most poignantly displayed when, following the captain's death a few years later, a ceremony was held at his hometown in Lewiston, New York:

> His coffin was covered with the star spangled banner and union jack, the flags which had for years floated over him, and as a token of respect to his memory, the flags of both nations were hoisted at half mast, from the time of his death until the earth closed upon his remains.[47]

The transnational nature of the tourist industry in the Niagara–Great Lakes Basin, and the interconnectedness of the New York canal system with its ties to Canada and the west were regularly commented upon by northern travelers. Indeed, the growth and prosperity of the New York canal system was partially grounded in the Canadian provinces. Travelers attributed much of Albany's growth to a flourishing trade with the internal states, and more convenient transportation and trade ties to Canada. During a stop there, one traveler wrote: "The situation of this city is advantageous, both from the direct communication which it enjoys with the Atlantic, by means of sloops and schooners, and the large tract of back country which it commands. A trade with Canada is established by means of the Erie and Hudson canal."[48] John Benwell described Albany in the late 1830s as the "depot for produce, especially wheat, brought via the Erie Canal from the interior; being in fact the storehouse of the trade to and from the interior States of the Union, the west, as well as from Canada and the Lakes."[49] Meanwhile, another traveler observed: "The object of the Erie Canal is to form a communication between New York and the internal, or Mediterranean seas of North America," and that of the "Champlain Canal is to form a communication through Lake Champlain, and the river Richelieu or Chambly, with the St. Lawrence and Canada."[50] Though the Champlain has received less scholarly attention than the Grand Erie Canal, this important northerly route similarly benefited from commercial and recreational ties to Canada and the St. Lawrence River.

Canada's impact on the New York Canal system was also revealed by the waves of emigrants at Albany who, alongside the northern traveler, boarded the Erie and Oswego Canals in transit to the northern provinces and the west. While on official business in 1830, Lord Durham rode the Erie Canal where he witnessed a "very considerable number" of emigrants "who enter the province by way of New York and the Erie Canal."[51] Colonial agents were also available in New York and Buffalo at the western terminus of the canal to assist the emigrants going to Canada. A well-known authority on the subject writes that settlers in route to Upper Canada were able to pass through the Erie Canal without paying duty on their luggage and goods, indicating the large wave of emigrants in route to the Canadian province, the profitability of this traffic in generating revenue on the Erie Canal, and the sense of good will between the neighboring countries generated by the canal age.[52] The *Canadian Emigrant and Western District Advertiser* also noted that emigrants landing in New York were now more readily using the Oswego Canal to cross into Canada from Lake Ontario, which contributed to larger numbers of settlers going north instead of remaining in the United States.[53] Such accounts offer a unique glimpse into Canada's integral role in bringing economic growth and development to New York State's canal system, while also illuminating the American canal systems in promoting Canadian settlement and westward expansion.

The topic of Canada became a regular source of entertainment as canal and stage drivers whiled away the time guiding visitors to the Falls. Traveler James Stuart recounted how an Erie Canal packet driver entertained passengers with stories of Canada's late War of 1812 hero, Sir Isaac Brock. "The canal agent spoke of him in terms of great respect, Stuart wrote, "as the best commander the British had ever sent to Canada" and that his death "was equally regretted on both sides."[54] The *Upper Canada Gazette* confirmed such sentiments, telling of groups of American visitors who, while clamoring to see the hero's monument at Niagara, lamented Brock's death "as much as they would have done that of any of their own generals."[55] Leaving Rochester for the Falls, one traveler overheard a conversation between two passengers about the curious practice of cross-border smuggling, and how a person could get rich through the illegal trade with Canada.[56] Others learned that "thousands of chests of tea were conveyed on the Erie Canal to Rochester and the Niagara River," and some people became "very rich in consequence" of this illicit trade.[57] Meanwhile, Rochester's famous Genesee aqueduct which would come to figure in the history of Canada's internal improvements development was also of interest to passengers. A Romanesque-like structure that spanned some 802 feet across the Erie Canal, the aqueduct was one the most impressive pieces of craftsmanship in North America.[58] Begun in

1821, the original contract was awarded to a Mr. William Britton but, as fate would have it, Britton died, and a new contract was granted to Alfred Hovey, a well-known canal builder from New York who, after completing the aqueduct in 1822, went to neighboring Canada to join a team of American contractors working on the formidable Welland Canal Deep Cut.

From Rochester, the northern tourist was encouraged to leave the canal and take the famous "Ridge Road" that went from Rochester to Lockport and the Niagara River where a ferry awaited passengers wishing to visit Upper Canada. Still in existence today, this ancient east-west thoroughfare across western New York acquired commercial and recreational importance during the canal age.[59] For those crossing the Ridge Road in route to Niagara during the Erie Canal's building, they would have been fascinated by the supplies of gunpowder, animals, workers and other provisions being transported along the road to Lockport where excavation was going forward on the much publicized Lockport Locks, an engineering marvel that became another anticipated attraction on the Northern Tour.[60] Indeed, few travelers crossed this route without commenting on its interest and utility. A travel diary entry in 1821 included the "Ridge Road" as a prominent feature of the Northern Tour, carefully detailing its uniformity and composition: "The stones in it appear to have been worn round and smooth by the surf; and it is interspersed with clam shells, and other substances" that resemble "the work of man."[61] Another traveler commented on "the well known Ridge-Road to Lockport," especially with its "proximity to the canal on one side, and Lake Ontario on the other" providing "facilities of intercourse with the great markets."[62] (See fig 6.3 on page 156.) Canadian William Lyon Mackenzie observed: "The ridge road is naturally good, perhaps the best in America," further indicating the tourist interest in internal improvements schemes. Having once reached Lockport, travelers were often "assailed by a host of canal and stage-agents and line-boat captains" eager to promote "their different modes of conveyance."[63] Energized by this new-found interest in canals and other internal improvements as curiosities in their own right, local hotelkeepers and entrepreneurs rushed to capitalize on the canal mania that was bringing flocks of travelers to the Niagara region.

No event in the history of Niagara Falls more clearly illuminates the link between tourist attractions and the transnational character of transportation and tourism in this region than one of the most bizarre and highly anticipated events in the history of Niagara Falls. In the summer of 1827 it was announced that a great lakes schooner called the *Michigan* "with live animals aboard" would be released over Niagara Falls. Thousands of

BRITISH NIAGARA FALLS.
Two Routes—by Coaches or Steam Boat—Take your Choice.

THE subscriber, proprietor of the PAVILION, at the British Niagara Falls, begs leave most respectfully to announce to the Ladies and Gentlemen, travelling to the Falls, that they can now choose their own mode of passing thither, by Coaches, or the Chippawa Steam Boat. The Boat leaves Buffalo, every day, at 1-2 past 8 o'clock, A. M. and returns from Chippawa, at 9 o'clock, P. M. Post Coaches leave the Eagle Tavern, at Buffalo, every day, at 1-2 past 8 o'clock, A. M. and passing through the village of Black Rock, cross the Niagara, by a safe and expeditious Horse Boat, and, if the Ladies and Gentlemen choose, they can visit *Fort Erie*, and view the *Battle Grounds*, without any additional charge; and arrive the at Falls, generally 1-2 past 11 o'clock, A. M.—Returning to the Eagle Tavern at Buffalo, at 7 o'clock, P. M. The price of fare, the same both ways. The Coaches and Horses, are the best west of Albany. The road is as smooth as a bowling green.—Extra Carriages furnished at all times on the shortest notice. No Drivers are employed, but those who are sober, intelligent, and attentive.

WM. FORSYTH.

August, 1826.   84

Figure 6.3. **Travel Options on the Northern Tour.** Part of the excitement of the Northern Tour was to diversify the travel itinerary and opt for the latest mode of land and water craft that the canal age had to offer. Courtesy of the *Black Rock Gazette*.

curious spectators dashed to the region to witness the spectacle. This bizarre attraction was organized by businessmen from both sides of the border. Two Canadian hotelkeepers, William Forsyth and John Brown, in association with American proprietor Parkhurst Whitney, sponsored the *Michigan*'s voyage. Months before, the trio signed their names to a broadside heralding the upcoming exhibition, and publicized it in newspapers throughout the United States and Canada. In addition to owning the area's largest hotels, Whitney and his Canadian partners dominated the transportation facilities around the Falls. Also benefiting from the events surrounding the *Michigan* was Augustus Porter, the elder brother of now famed General Peter B. Porter, who not only monopolized the region's mercantile and transportation industry in the Niagara–Great Lakes Basin but also dominated a large share of the tourist industry.[64]

The men responsible for hosting the *Michigan* had spent much of their lives promoting internal improvements, commerce and tourism along the Niagara–Great Lakes borderland. At the forefront of the exhibition was Canadian businessman William Forsyth who was among the most interesting and colorful settlers in the region. His father, a loyalist farmer, came to Upper Canada in 1783 when William was still in his teens. William initially followed in his father's steps as a yeoman farmer, but his loyalty to King and Crown was tested in 1799 when, following a criminal offense, he tried unsuccessfully to escape across the border to the United States. Later, Forsyth used his familiarity with the Niagara River to smuggle goods between the United States and Canada. At the end of the War of 1812, Forsyth became interested in the tourist business. By 1822 he opened the doors of the impressive three-story *Pavillion* Hotel with its expansive balconies and piazzas overlooking the Horseshoe Falls. Next to John Brown's prestigious "Ontario House" that was also adjacent to the Falls, Forsyth's *Pavillion* retained the distinction as one of the most elegant hotels in North America. In addition to accommodating cross-border tourism and travel, Forsyth promoted other improvements that would allow Upper Canadians to more easily conduct their business in the neighboring Republic. In 1834, he offered to operate an additional ferry service above Niagara Falls so that residents from the Canadian side could have their "grinding, carding, fulling, etc." done in Black Rock, New York's extensive mills[65]

Forsyth and Brown were well acquainted with Parkhurst Whitney who operated the popular *Eagle Hotel* on the American side of the Falls. The demands of the tourist trade along the Niagara borderland encouraged these large proprietors to unite their energies and promote commerce and transportation across the river. Forsyth and Whitney had joined forces earlier to start the first regular ferry service below Niagara Falls. So when the huge crowds descended upon the region in 1827 to witness the *Michigan*'s

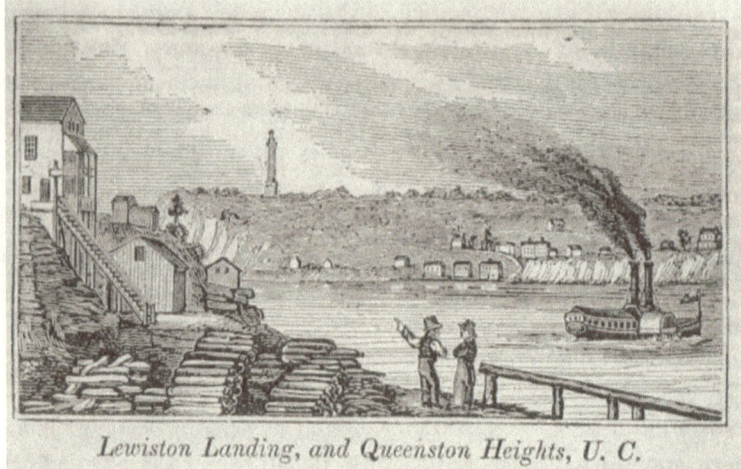

Lewiston Landing, and Queenston Heights, U. C.

Figure 6.4. **Lewiston, NY, and Queenston, Upper Canada, Ferry Landings.** As a result of canal and transportation innovations at Niagara, ferries and steamboats regularly plied between the Canadian and American sides. John W. Barber and Henry Howe, *Historical Collections of the State of New York*. Image courtesy of the Library of Congress.

plunge, they arrived on ferries, steamboats and coaches owned and operated by the same enterprising leaders who had promoted the event (fig. 6.4).[66]

One of Niagara's oldest settlers, American entrepreneur and tourist developer Augustus Porter, well understood the attraction of internal improvements in bringing visitors to the region and prosperity to the borderland economy. In fact, the most coveted vantage point for viewing the *Michigan*'s perilous descent was the well-known Goat Island Bridge that Augustus had erected a few years earlier. Acting as his own engineer, Augustus built a bridge between the main American shore and the island that was able to withstand Niagara's famous rapids above the Falls. During his visit to Niagara, Captain Basil Hall seemed as fascinated by Porter's bridge as by the Falls, calling it "one of the most singular pieces of engineering in the world," that "shows not only much skill and ingenuity, but boldness of thought in its projector."[67] Situated no more than "fifty yards above the crest of the Falls," Hull concluded that "without the assistance of drawings . . . it would scarcely be possible to render intelligible any account of this extraordinary work, which has added much to the interest of Niagara."

As the region's most prominent resident and developer, Augustus promoted the tourist trade at Niagara. He also hosted a number of distinguished guests at his spectacular home overlooking Niagara Falls, including President John Quincy Adams, General Lafayette, Charles Dickens, De Witt Clinton, and Andrew Jackson's arch-nemesis and notorious national banker Nicholas Biddle, president of the Second Bank of the United States. Biddle, like Augustus, was a keen friend of internal improvements, and at Augustus's bequest, he made a special visit to the Falls in 1827 where he was welcomed by Augustus's renowned brother, Peter, who had just completed his term as Secretary of War under John Quincy Adams. A story is told that Biddle was standing over the Falls when he spotted a stairway on the Canadian side built "by the subscriptions of certain citizens of Boston." Perceiving "the increased interest which would be added to the scenery," Biddle underwrote the construction of a similar staircase on the American side that was fittingly named *Biddle's Staircase*.[68] It was an eighty-foot-tall enclosed wooden structure with tiny portals that offered a panoramic view of the Canadian and American falls. Access to the stairway was free, making it one of the most popular improvements at Niagara during the canal age.[69] The mutual efforts of Canadian and American entrepreneurs and businessmen to make the Niagara borderland more accessible and friendly to visitors ensured a tremendous international turnout of spectator's for the *Michigan*'s grand descent over the Falls.

Scholars generally interpret the *Michigan* as the first of many spectator stunts at Niagara Falls that forever transformed the nature and sublimity of this majestic wonder. As one scholar writes, the *Michigan* represented "the moment that crass materialism began replacing natural wonder" at Niagara Falls.[70] In her study of Niagara Falls, Elizabeth McKinsey notes that the *Michigan* ushered in a new type of "consumer attitude" toward the Cataract that forever changed the face of the Falls. She attributes this change to the opening of the Erie Canal in 1825 that assured the exploitation and spoliation of the falls, noting: "a national icon, remote and imaginistic, was now transformed into a fashionable resort, accessible by easy transportation." With the arrival of some 15,000 visitors to the Falls during the day of the *Michigan*'s launching "came all the tourist accommodations and allurements: hotels, guides hawking carriages or oil-cloth clothing, souvenir stands, museums, specialty-created vantage points . . . and crowds." The result was "a transformation of the image of a wondrous natural prodigy into a public spectacle."[71]

The *Michigan* held deeper meaning for its promoters, and for many of the tourists' witnessing its grotesque end. While the *Michigan* certainly did contribute to the crass commercialization associated with the falls, in another sense the event was deliberately calculated by the Canadian-American boosters to project and symbolize the importance of transportation, commerce, and tourism in strengthening the borderland community. Even more, in less obvious but surprising ways, the symbolism of the *Michigan* served as the strongest confirmation of how canals and other internal improvements had redefined the meaning of the international boundary and the physical boundary that Niagara Falls epitomized so dramatically.

Journalists covering the *Michigan* noticed the new type of popular entertainment culture that greeted the thousands of persons drawn by the *Michigan* spectacle. As told by Niles:

> Amusements, in all their rounds, were to be seen at all the public houses, and even by the wayside. There was Mr. Nichols, with his ventriloquial powers, teaching ladies the secret of talking without using their own tongues, . . . The dog Apollo, too was there, and played cards with, and gave lectures on astronomy to ladies and gentlemen without number. Grosser amusements for vulgar tastes, also abounded, such as caravans, mountebanks, learned pigs, etc. . . .[72]

Accommodations and public houses were so crowded that Cary of the *New York Observer* counted some "fourteen asleep *upon*, and thirty-six *under* the billiard table" at one hotel.[73] "The public houses at the Falls were so thronged," wrote another, "that almost every inch of the floor was occupied as comfortable sleeping apartments."[74] As an event, the releasing of the schooner *Michigan* over Niagara Falls exceeded even the promoter's wildest expectations.

But the *Michigan* also made a more serious statement, clearly legible in its context about improvements reshaping the borderland, and the larger North American transportation revolution that brought economic and recereational prosperity to it. By the time of the *Michigan*'s historic launching over Niagara Falls, the Erie Canal had been successfully operating for two years, and the much-anticipated Oswego and Welland Canals were near completion. As anyone connected to transportation, trade, and travel understood, these improvements allowed shipping and commerce to connect what the barrier of Niagara had kept separate in nature, and in the course *overcoming Niagara*. Linking the two by dramatizing a schooner destroyed by that very barrier thus provided a taunting, ironic way to celebrate its defeat through canals and other internal improvements. In this way, while promoting tourism at Niagara, the *Michigan* boosters heralded the promises of the North American canal age, and the cross-border community and friendship it was nourishing along the Niagara River and throughout the borderland.

By way of making this linkage explicit, the Erie Canal's celebratory opening was deliberately echoed in the ceremonial staging of the *Michigan*'s plunge over Niagara Falls in 1827. As recounted in an earlier chapter, Governor De Witt Clinton's famed *Seneca Chief* traveled the Erie Canal from Buffalo to New York City in 1825 for the state-wide ceremonial opening. The *Seneca Chief* was trailed by another barge called the *Noah's Arc* and its replacement at Rochester, the *Young Lion of the West*.[75] According to Cadwallader Colden's *Memoir* (1825) of the Erie celebration, these vessels "had on board a bear, two eagles, two fawns, with a variety of other animals . . . *all products of the west*."[76] Not coincidentally, the Niagara boosters drew from this language and symbolic pageantry as seen in the broadside announcing the *Michigan*'s fateful voyage over Niagara Falls: "The greatest exertions are making," the broadside proclaimed, "to procure animals of the most ferocious kind, such as Panthers, Wild Cats and Wolves; but in lieu of some of these . . . a few worthless or vicious Dogs . . . and perhaps a few of the toughest of the Lesser animals will be added to, and compose the cargo." The Niagara boosters also advertised the *Michigan* as a "Grand Aquatic Display" to be performed over the Falls, imitating the language of the Erie Canal's ceremonial performance that was dubbed a "Grand Aquatic Display" between Buffalo and New York City two years earlier.

Several weeks before the *Michigan*'s sendoff, an advertisement in the *Black Rock Gazette* announced that the "Sch. Michigan, with all her tackle, will be sold at public auction."[77] (See fig. 6.5.) Everything but the hull was sold, including the sails which were purchased by a Canadian Captain. Major Donald Fraser, a resident of Black Rock and a personal and business ally of Porter, Barton and Company, purchased the *Michigan* with the sole intention of sending it over Niagara Falls.[78] Fraser was closely connected to Augustus and Peter Porter through personal and mercantile dealings on the Great Lakes. So attached was he to General Porter that, upon the latter's death in 1844, Fraser was honored with Porter's "best pair of pistols" used during the War of 1812.[79] Meanwhile, Capt. James Rough of Black Rock, *the oldest navigator of the Upper Lakes*, and a highly esteemed captain who sailed on the Great Lakes since before the War of 1812, was appointed to manage the enterprise. A day or two before the exhibition, several of Black Rock's citizens placed a *live caged buffalo* on board the vessel, brought all the way from the Rocky Mountains, in addition to a bear brought from Canada. The staging of the buffalo may have been a way of poking good-humored fun at the town of Buffalo that was still gloating over winning the prize of the Erie Canal's western terminus over Porter's hometown of Black Rock a few miles down the Niagara River.

If the *Seneca Chief* was the centerpiece of the Erie Canal's "Grand Aquatic Display" in the 1825 inaugural journey from Buffalo to New York City, the 132-ton schooner *Michigan* occupied a similar place in the identically named "Grand Aquatic Display over Niagara Falls two years later.[80] Just like the earlier Erie Canal celebration, bands, parades, flags, and music also accompanied the *Michigan* and her entourage on the day of send-off, along with a flotilla of steamboats, lake vessels, canal boats, and barges, all symbols of the canal age. The broadside's detail made the connection between the great lakes, canal era commerce, and the spectacle more explicit still:

> The first passage of a vessel of the largest class which sails on Erie and the Upper Lakes, through the Great Rapids, and over the stupendous precipice at Niagara Falls, is proposed to effect, on the 8th of September next. The *Michigan* has long braved the billows of Erie as a merchant vessel but having been *condemned* by her late owners as unfit to sail longer proudly "above;" her present proprietors, in conjunction with several spirited public friends, have appointed her to convey a cargo of Living Animals of the Forests, which encompass the great upper lakes, through the white tossing deep rolling rapids of the Niagara, and down its grand precipice, into the basin *below* . . .
>
> It is intended to have the *Michigan* fitted up in the style in which she is to make her splendid but perilous descent, at *Black Rock*, where she now lies. . . . The animals will be caged or otherwise secured and placed on board the "condemned Vessel. . . ." Such as may survive, and be retaken, will be sent to the Museums at New York and Montreal, and some perhaps to London.

The above description speaks clearly to the North American transportation revolution along the Niagara borderland. The broadside evokes an image of a *vessel of the largest class* descending Niagara Falls and in its course connecting *Erie and the Upper Lakes* with the sister lakes *below*. Embracing a continent-wide vision of commerce and trade, the broadside prophesized that once the *Michigan* descended the Falls *to the water in the Gulf beneath*, her cargo would inevitably find a market in *the museums of New York and Montreal, and some perhaps to London.* Whether goods were transported through the Erie, or soon to open Oswego or Welland ship canals, Americans and Canadians had much to look forward to from the opening of more markets and routes, regardless of the international border. As a further symbol of cross-border unity, a British Union Jack was ceremoniously hoisted above the *Michigan* alongside the Stars and Stripes before

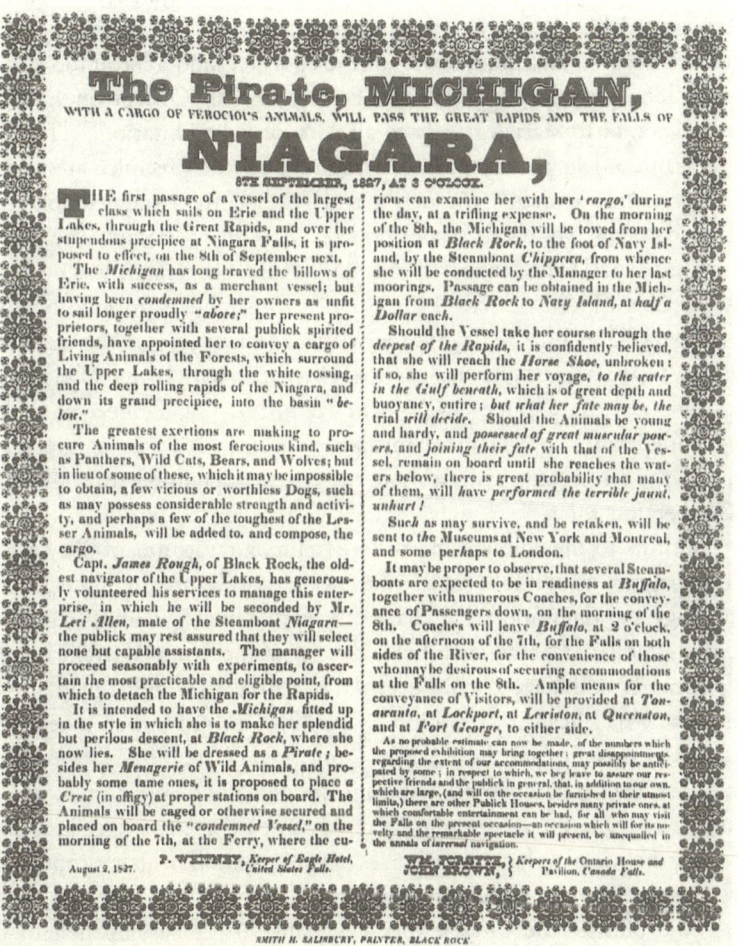

Figure 6.5. *Michigan* **Broadside.** Courtesy of the Buffalo History Museum.

send-off. For those connected to transportation, trade and tourism along the Canada–United States borderland, the message of change wrought by improvements and canals was clear within a little more than a decade after the end of the War of 1812: cooperation and friendly competition, rather than conflict and rivalry, went hand in hand.

Niagara's hotelkeepers began preparing weeks in advance of the *Michigan*'s send-off. John Brown of the *Ontario House* in Niagara, Upper Canada, joined the Buffalo *Mansion House* in offering a daily line of coaches between their establishments, and for the convenience of their patrons, providing extra coaches at both locations.[81] William Forsyth of the popular *Pavillion Hotel* assured his guests that regular stages ran "daily to Buffalo, Niagara, Queenston, Lewiston, and all parts of the country," and a "steamboat ran all summer long between Chippewa and Buffalo." From "Queenston, Lewiston, and Niagara, boats start regularly to all parts of Lake Ontario."[82] "The roads are pleasant and in good repair. The stairs down the precipice are safe and the ferry well attended."[83] Augustus and Peter Porter arranged for several of their company vessels to accommodate the swell of visitors. The growing transnational and interconnected character of the borderland was evident in how the *Gore Gazette* described the parade of steamboats from both sides of the Niagara River on the day of the *Michigan*'s launching: "the steamboats Canada, Niagara and Queenston; the Ontario and Martha Ogden, from Lake Ontario; the Chippewa, William Penn, Niagara, Pioneer and Henry Clay from Lake Erie, the three former laden with subjects of the sea-girt isle, and the seven latter with a more numerous deputation from the family of Brother Jonathan."[84]

Visitors to the Falls during the weekend extravaganza would have had an opportunity to witness some of the boldest advances in transportation technology and innovation that characterized the canal age in this region. The recently established horse-ferry boat, the *Cossack*, offered constant crossings for passengers and stagecoaches between the American and Canadian sides. The *Cossack* had been in operation successfully for more than a year above the Falls between Black Rock, New York, and Waterloo, Upper Canada. It met a growing demand from businesses and populations on both sides of the river. One of the most novel marvels of the canal age, in place of oarsmen the *Cossack* had four horses on a treadmill which propelled the ferry across the river in an astonishingly short time. By the time of the *Michigan*'s plunge, the *Cossack* was promising to get passengers across the river in five minutes—"some trips have been made short of 5"—a speed that "could not be equaled by any boat of a similar description in this state."[85] Other improvements like the "6-horse coach, the first ever used in the country," were able "to carry 20 persons . . . back and forth to the Falls in "modern style."[86] American journalist Hezekiah Niles marveled at the "vehicles of every possible description" witnessed during the *Michigan* extravaganza "from the John Bull coach and six, with its thirty passengers, down to the Canadian pony, rode by two Indians, either of whom looked better able to *carry*, than to be carried by the beast."

Saturday, September 8, was the designated day of the *Michigan*'s send-off. As stated by one account: "The stages came crowded, as well as the canal boats, so much so that it was difficult to find a conveyance to the Falls."[87] Visitors who came in anticipation of the much-heralded event were blessed with the same favorable weather conditions that welcomed De Witt Clinton and his flotilla in the Buffalo harbor in 1825. According to one *Michigan* spectator: "The sun shone in full splendor, the waters of the Erie were placid," there being scarcely a ruffle upon its surface. . . ." Another observed that "the day was as favorable as could have been expected."[88] The *Michigan* was stationed at the Black Rock harbor where "the curious could examine her with her *cargo* . . . at a trifling expense." Augustus Porter prevailed upon his brother Peter, who two years earlier participated in the Erie Canal celebration, to make sure he arrived with his family in time to see the much-anticipated spectacle, writing "I think it will be most interesting to see the great commotion of people at the Launch. You must not fail to come."[89] The intensity of the occasion was described by another spectator: "Wagons, filled with country people rattled through the town all night, and on Saturday morning Buffalo itself seemed to be moving in a mass towards the grand point of attraction."[90] At 3 p.m., the *Michigan* would be towed by the jointly owned Canadian and American steamboat *Chippewa* to the foot of Navy Island where she would be released from "her last moorings."

The Niagara boosters were more than gratified by the thousands of people who came to witness the event. By one o'clock, swarms of people had descended upon the Falls, positioning themselves on the banks of the river, on the islands, the housetops, balconies, and treetops above and below the Cataract. Every "place and every corner and nook was filled with human beings," wrote William Lyon MacKenzie, "bands of music enlivened the scene—and the roar of the African Lion in the menagerie . . . were almost too much for human organs."[91] American editor Niles traveled a long distance to get to the Falls in time for the *Michigan*, musing that he found himself engulfed by a crowd who all "seemed alike indifferent to everything but to see the ship go over the falls."[92]

Visitors actually paid to ride alongside the *Michigan* and her cargo of live animals before her release above the Falls. Steamboats and barges, filled to overflowing with thrill-seekers, accompanied the doomed vessel to a place where she was anchored within sight of the Falls. Passage aboard the *Michigan* itself, from Black Rock to the foot of Navy Island could be obtained for 50 cents. Though it never transpired, rumor had it that a *human daredevil* would pilot the *Michigan* to the brink of the Falls, and just seconds before reaching the abyss, escape in an air balloon.[93] A more bizarre account told of a man who was to be executed at Niagara, Upper Canada, the following Monday, and thus "earnestly begged of the court the *favour* of steering the condemned ship over the Falls of Niagara."[94] Seeing how visitors relished watching animals or maybe even a human being go over the Falls to their likely death inspired local innkeepers. Two years later they invited notorious stunt man Sam Patch to perform his dramatic leap over the Canadian Falls![95]

At a quarter to three, the *Michigan*, with her cargo of helpless animals, made an appearance a mile or so above the falls. The steamboat *Chippewa* towed the *Michigan* as close to the Falls as possible before turning the condemned vessel over to a barge manned by Captain James Rough and several stout oarsmen. This was the moment of most intense excitement. It was claimed that if Captain Rough had waited any longer to cut the rope "he would have continued to tow the *Michigan* until it would have been impossible for them to escape."[96] According to *Niles*, "the interest felt for the safety of the barge was painful. Human life had never before, in that frightful manner, been voluntarily periled."[97] "Indeed," wrote another observer, "such was the fear of the hands, . . . that on approaching near the rapids they cut the tow-line before they had received orders from their commander."[98]

The *Michigan*, which had "long braved the billows of Erie with success, as a merchant vessel," was now abandoned to the mighty rush of the Cataract. An eyewitness account dramatizes the events that followed:

> Every eye which could command a view of it, was riveted on the Michigan at this moment—and when she made the first plunge into the rapids, there was a simultaneous shout of applause—the shock was evidently a severe one, and its effect was visible upon her heterogeneous *ship's crew*, which now began to bestir themselves—his buffaloship was evidently in uneasy quarters—the eagle vainly essayed to soar from the troubled waters. . . . Before arriving at the second set of rapids, the vessel stuck apparently

between two rocks, for a few seconds, but the violence of the current drove her round, and she went stern foremost over the second ledge—pitched on her starboard side, and before righting, both her masts were carried away. The Buffalo and several other animals were thrown overboard. . . . After this shock the vessel became waterlogged, and floating down the rapids, exhibited successively a wreck on the breakers, and a ship going to pieces. She broke right in two halves, and meeting with no further obstruction, darted like an arrow down the fearful steep.[99]

The spectacle, "from the moment she hove in sight until she was dashed into, literally a thousand pieces, was full of interest and novelty." In a last minute rush, the crowd ran below the Falls where pieces of the ships wreckage were eagerly salvaged for souvenirs. "Such was the eagerness of the multitude present to procure a piece of her that before sunset a great part of her was carried away." Most of the animals were killed or escaped before going over the Cataract. One animal to survive the Falls—a badly hurt goose with its head cut and eye poked-out—was taken by the cheering crowd and put on display at a local museum.[100]

The unbound appeal of the *Michigan* spectacle was evident in the masses of people who came from all parts of the western world to witness the condemned vessel go over the Falls. Several noteworthy guests could be found among the crowd of thrill-seekers, journalists, and curious bystanders.[101] These included Secretary Montoya of the Mexican Legation who was one of the first representatives to come to the United States since declaring Mexican Independence.[102] Canadian William Lyon McKenzie, editor of the *Colonial Advocate*, enthusiastically recorded the day's events. Widely known for his muckraking and virulent opposition to the Upper Canadian rulers, McKenzie was a champion of Jacksonian democracy and a friend of the neighboring Republic. Ironically, a decade or so after the launching of the *Michigan*, McKenzie, during an assault on the Upper Canadian establishment, would see his own vessel, the *Caroline*, descend the Falls in flames.[103] Other visitors who witnessed the *Michigan* included laborers from the Welland Canal, it being noted that construction on the line was delayed because the exhibition "drew the attention of many of the hands from the work."[104] The presence of the Welland Canal laborers at the Niagara spectacle suggests that tourism and leisure were becoming accessible to working-class people during the North American canal era. Such a "mingling of spirits, ages, and sexes," wrote Niles, "such a mass of heterogeneous humanity," had never before been witnessed at the Falls.[105]

The *Michigan* awakened visitors to the larger North American transportation revolution. In the days, months, and years following the *Michigan*'s ill-fated voyage over the Falls, ever more tourists came to the region to witness the canal boom and other internal improvements attractions associated with the age. The spectacular success of the *Michigan* symbolized and solidified trans-border ties in the Niagara–Great Lakes region, confirming the belief among Canadian and American business developers and entrepreneurs that their shared border must be kept open and free of restrictive impediments. Advances in land and water transportation ensured an even greater flow of peoples, goods, capital, and technology across the porous border while furthering interregional and transnational linkages in the Niagara–Great Lakes Basin.

Events like the *Michigan* also encouraged hotelkeepers and businessmen to promote exclusive two- to three-day tours of the major canal towns and sites on both sides of the Niagara border. The newly advertised canal and water excursion commenced at Buffalo, "crossing the Niagara River at Black Rock in the Horse Boat, and branching off from Chippewa" a flourishing canal village at the southern end of the Welland Canal before "re-crossing the Niagara in the new boat at Youngstown and again diverging at Manchester and Lockport" before returning to Buffalo. Other tourists enjoyed riding the new ferry boats "worked by horses" that were becoming objects of curiosity in themselves.[106] Hotel operators along the frontier spent considerable money improving their establishments for those wishing to partake of the canal tour and the surrounding attractions. The *Black Rock House* offered fine accommodations "in the immediate vicinity of the Ferry, and of the steam and Canal Packet Wharves." One traveler, from the moment he checked into a hotel at Buffalo, had "a fine opportunity of observing the business of the canal and harbor, as it was commencing with the early morning. Such a spectacle of hurry, of sailing, and riding, walking to and fro, I have seldom if ever witnessed."[107] In her retrospect of western travel, Harriet Martineau partook of the local tour, noting that "the beginning of the Welland Canal was next pointed out to us," but because of inclement weather, "it was not a moment to care for canals."[108] While traditional attractions like the majestic Cataract and War of 1812 battle sites were still interesting, during the canal age it was man-made wonders like the Lockport Locks and the Welland Deep Cut that attracted so much attention.

Designed by New York native, Nathan Roberts, the Lockport Locks consisted of a flight of combined locks that carried boats over the sixty-foot rise in the Erie Canal.[109] Carved out of the Niagara escarpment, these giant stair-like locks were the crowning achievement of Robert's career, making him one of the most sought-after engineers in the United States and Canada. Lockport was now placed on the map, resulting in a flood of visitors who were as eager to see these "noble locks" as they were the great falls at Niagara. One of Lockport's most distinguished guests, the Marquis de Lafayette, came to christen the locks in June of 1825, thinking them as magnificent a man-made wonder as the incredible natural wonder of Niagara Falls.[110] According to a fellow tourist in 1828, "[n]o spectacle on the canal, however, impressed me so much, as Lockport. . . . You look up an elevation of more than 100 feet . . . nor can you realize, that any human power would raise a loaded canal boat to the dizzy eminence above. But the achievement is accomplished in a few minutes by the simplest means; and you find yourself raised to the level of the neat village above. The steps, the abutments, and all the lock work is firmly done in massive stone." Words could not describe the spectacle at Lockport, instead "you must see it."[111]

Many travelers agreed that the workmanship and mastery of the Lockport Locks made them an object as noble and interesting as Niagara Falls (fig. 6.6). Arriving at Lockport at daybreak, Horace Franz, Esq. declared the works in the Mountain Ridge and the Locks themselves "the great wonder of the canal." Franz believed the five double locks of stone a fitting monument to Clinton and the canal's projectors. Another traveler who came to view the famous five-tiered set of locks wrote: "Unmoved as I usually am by surrounding objects, I am willing to confess that I was more astonished than I ever was by anything I had before witnessed."[112] Visitors were in awe of the fact that "the canal is carried over what is called the mountain ridge, to the height of sixty feet, by a double set of five locks. One set is used for a boat ascending and the other for those descending. The whole of this stupendous work is finished in the most durable and scientific manner; and the scene from the number of boats passing through the locks is extremely animated."[113] Yet another traveler marveled at the five double locks "which elevate us 60 feet above the escarpment, we are therefore 560 feet above the Hudson and have attained the same elevation as the Falls."[114]

Travel guide guru Gideon Davison encouraged visitors to stop at the "magnificent locks," alerting them to the "finest imaginable workmanship, with stone steps in the center and on either side, guarded with iron railings, for the convenience and safety of passengers."[115] Taking Davison's advice, Canadian William Lyon Mackenzie planted himself on a large grey stone above the Locks and "surveyed the scene around me—the canal, the locks—the bustle and activity—wagons with ox-teams and horse-teams."[116]

Visitors agreed that the locks "furnish a noble specimen of the ingenuity, art, enterprise, and perseverance of man." The sense of sublime was not limited to experiencing the natural wonder of Niagara Falls. On going through the locks, one visitor wrote: "Lighted by lamps, there is much sublimity attending the passage through them in a dark night."[117] Famed visitor Nathaniel Hawthorne overheard two Michigan traders say that while on the whole they thought Niagara Falls worth looking at, "they would go twice as far to see the noble stone works of Lockport, where the Grand Canal is locked down a descent of sixty feet."[118] For the natural scientist and geologist an added bonus was "the fine opportunity of seeing these older fossiliferous (sic) rocks laid open to view" by the Lockport excavations, again illuminating the canal's role in bringing education and science to many.[119]

The transnational nature of Niagara's growing tourist industry was evident in the number of travelers who, after seeing Lockport and the Falls, went to witness Canada's Welland Canal with its massive Deep Cut. In his much read *Northern Traveller* (1830), Theodore Dwight urged visitors at Niagara Falls to devote some time to the Welland Canal, "this new, important and highly interesting work." Dwight was astonished by the magnitude of the Canadian project which had managed to overcome the obstacle of Niagara Falls: "The largest cataract in the world, which presents a scene of confusion, uproar and tumult, . . . was to be surmounted by a system of works in which the rush of the element, so awful, resistless, and destructive, was to be curbed and tamed to a calm and gentle descent, and rendered useful to the objects of commerce."[120] Basil Hall was no less enraptured by the Canadian work:

> The Welland Canal is intended to perform the same step over the intervening land, as that made by the Falls and Rapids of Niagara, from the level of Lake Erie to that of Lake Ontario— only in a more gentle and manageable way. Niagara is wonderful and beautiful to look at, and so far has its advantages. But this great canal will be much more useful in advancing the ordinary business of life. The feeling excited by seeing these two noble works, the one of nature, the other of art, side by side, at a distance of only six or eight miles, are very different I grant; but both in their respective ways, are well calculated to fill the spectator with agreeable reflections.[121]

The Welland Deep Cut, an almost two-mile stretch of the canal that was dug through the solid rock of the Niagara escarpment, was of particular interest to travelers and canal buffs. Similar in scope to the Erie Canal deep cutting at Lockport, the Welland cut was evidence of incredible engineering prowess, attracting gifted canal builders like Nathan Roberts whose experience at Lockport made him a most desirable employee of the Welland Canal Company. Begun in September of 1825, the Canadian Deep Cut, like its American counterpart, was one of the most herculean undertakings in the history of canal construction in North America. So formidable was the work that, after surveying and estimating the cut, Roberts warned in an 1824 letter to the Welland Company that from his own experience at Lockport, "and from the unforeseen difficulties attending the same, I have been induced to give such an estimate, that you may not be deceived in its accomplishment, for it is almost impossible for any engineer to estimate this description of cutting without having witnessed the operation."[122] As time would tell, Roberts's list of difficulties and unforeseen problems proved astutely accurate.

The challenges posed by the Deep Cut were recorded by those closest to the project, and by the flood of correspondents and curious spectators who came to visit the site. According to one witness, at the very commencement

Figure 6.6. **Entrance to Lockport on the Erie Canal.** Travelers to Niagara were as fascinated by innovations in canals and other internal improvements as they were by the sublimity of the Falls. The Miriam and Ira D. Wallach Division of Art, Prints, and Photographs: Print Collection, the New York Public Library. "Entrance to the harbor, Lockport," courtesy of the New York Public Library Digital Collections.

of the work "a well was dug to the foundation" and the pit filled with water so quickly that the workmen "were obliged to retreat as fast as possible." Merritt described the formidable nature of the undertaking, noting that the entire distance through the cut was "one mile 54 chains, averaging about 44 feet cutting. The depth of from 12 to 18 feet from the surface, it is composed of clay, with a small mixture of sand, and below this—tenacious blue clay." The project's primary overseer, American builder Oliver Phelps, likewise recalled the difficulty of the task, writing that the soil was hard "there were many streaks of hard pan that could not be ploughed, but had to be picked—there were many large stone and rocks which took from four to twelve yolk of oxen to draw out the ditch—there were some so large they had to be blasted."[123] By the project's end, some 1,477,700 cubic yards of earth were excavated from the Deep Cut with only the crudest of tools and instruments. This mass achievement was praised in an 1828 edition of the *American Journal of Science and Arts*:

> The bottom is removed from each end of the cut with scows, and the earth is deposited in the Welland river and in a large reservoir below bottom level at the other end. Between these points, the earth is removed with carts, wagons, and machinery; being drawn to the top, where it is deposited on the bank, on either side. The machine in most general use, is a common wagon wheel, fixed on an upright post, about seven feet from the ground on the top of the bank; a rope with a hook on each end reaching from the bottom of the canal to the top, is fixed round this wheel which hooks on the back of the descending cart and to the tongue of the one below, so that the return team assists in pulling up the loaded one, thereby, in effect, reducing the ascent to a perfect level, as the loads are drawn up with more ease than they are removed on the level of the discharge.[124]

Englishmen Joseph Pickering made an excursion to the Deep Cut in 1827 where work was still progressing, noting that "several hundreds of people were excavating the canal; hands are in request to get it through as soon as possible." Pickering was struck by the labor-intensive nature of the project observing, work is "from sunrise to sunset" but remained confident that "[t]he Welland Canal will unlock the energies of the west—the granary of America."[125] Pickering went on to comment that "the inhabitants are chiefly Americans," indicating the prevailing influence of the United States in the building of the Welland Canal. The flurry of activity on the Deep Cut was observed by another spectator who saw "5 or 600 men, with 500 oxen" employed on the work. Stories were told that the cattle and oxen in the canal bed were so exhausted from pounding the earth that they were hardly fit for consumption. The dangerous nature of work on the Deep Cut was evident in other accounts of workers and animals "dying off by scores."[126]

In order to accommodate the large flow of people wishing to observe the Welland's progress, several enterprising New York businessmen began promoting recreational trips to the neighboring canal. In 1827, the Canadian *Farmer's Journal and Welland Canal Intelligencer* notified the public that the American packet *Experiment* "was lately brought from the Erie Canal, by a number of spirited countrymen, for the accommodation of pleasure parties on the canal between the [Welland Canal] harbor and St. Catharines now completed."[127] Visitors were in awe of this colossal Canadian work. All those who visited the Deep Cut, chimed one American newspaper, "could not have but admired the enterprise which undertook so mighty an excavation."[128] Travel expert Theodore Dwight thought it "the greatest work of the kind in America," and tourist James Stuart agreed that the "Deep Cut," *par excellence*, was "the greatest artificial work in America. . . ."[129] No other work equaled the "stupendous deep cut," wrote one Yankee admirer, concluding that "few if any works of art yet constructed in America" were "more worthy of a visit."[130] *Mitchell's Compendium* on internal improvements described Canada's Welland Canal as "one of the more important works. . . . The deep cut, and locks are considered one of the most striking spectacles in North America." Such embellished accounts indicate that the Welland Deep Cut and locks, just like the Lockport Locks, had become as great an attraction as the Falls of Niagara. *The American Traveller* agreed that in addition to the Falls, "The Welland Canal" is an interesting object "which deserves attention."[131] Even today, hundreds of American tourists annually flock to the Welland Canal where they can see up-close the wonder of the world-renowned Welland lock system. Thanks to developments on both the Welland and New York canal systems, tourism on a grand scale acted as a catalyst to commercial and recreational expansion at Niagara and the surrounding borderland vicinity.

The transnational nature of the tourist industry in the Niagara–Great Lakes Basin made the United States and Canada each other's greatest admirers and friends. Canals and other internal improvements helped overcome natural and artificial barriers as manifested in the development of one of the most important interlocking, interconnecting navigation systems in this region. The north-south orientation of canals and other internal improvements had created a porous border that promoted commercial expansion and settlement, tourism and progress in general between the historically divided peoples. An 1837 ode entitled "Niagara to its Visitors" celebrated these developments, and put them at the center of the Niagara Falls story, neatly encapsulating an important theme advanced in this chapter and book:

> Oh ye, who come from distant climes,
> To visit me and read my rhymes, . . .
> Through Lake Superior, it is true,
> I descend from old St. Louis . . .
> Michigan nurses me in her lap,
> Huron feeds with Saginaw pap, . . .
> Through Erie next I guide my stream,
> And learn the power and use of steam.
> Although the rapids rather flurry me,
> And into wheeling whirlpools hurry me,
> The Devil's Hole does most me scare, I oh!
> And makes me glad to reach Ontario.
> Traveled so far 'tis thought of vital,
> Importance I should change my title,
> And though it should be his abhorrence,
> They make my sponsor old St. Lawrence.
> The course I steer is rather critical,
> For not much liking rows political,
> 'Twixt' both my favors I divide—
> Yankee and British on each side.[132]

Summary

For Captain Basil Hall, who saw many "interesting objects in the neighborhood" of Niagara Falls, "the most amusing were a trip to Buffalo . . . where the great New York Canal commences—and a visit to the Welland Canal, which joins Lake Erie with Lake Ontario."[133] Receiving international acclaim, visitors traveled from afar just to experience the progress in land and water craft or to get a glimpse of the latest canal and engineering technology that heralded the canal age in this region. It was this same enthusiasm for canals and other internal improvements that helped solidify the borderland economy as Canadian and American business leaders, hotel operators, transportation innovators, and entrepreneurs mutually promoted the tourist industry and the free flow of people, goods, commerce, technology, and information across their shared boundary.

If at the more distant centers of power, the nineteenth-century canal age seemed like a game of tug-of-war for political and commercial control of the continent's interior, along the Niagara borderland it was shared Canadian and American experiences and interests, rather than conflict and rivalry, that best defined the transportation revolution in this region. Nowhere was this more dramatically symbolized than in the staging of the *Michigan* that celebrated the internationalization and globalization of transportation, trade and commerce in the borderland region, its triumph over the obstacles of nature and geography, and Niagara's rising distinction as a premier tourist destination in North America even as this was based on one of the world's most spectacular natural attractions. The power and appeal of the transformations wrought by the canal age was most succinctly stated by one northern traveler who, after a customary stop at Niagara's majestic Falls, wrote that he feared being "excommunicated by my American readers" for visiting "neither the Erie nor the Welland Canals; nor even the locks at Lockport, or the Deep Cut." The same visitor made sure to congratulate the spirited individuals who undertook the Welland Canal which "by its means, the obstacle presented by the Falls of the Niagara, have been effectually overcome. . . ."[134] National rivalries and competition had been stemmed as a result of the complementary New York and Welland Canal systems, leaving in its place a more porous and permeable border.

7

Conclusion

> The time may come when steamboats up
> Niagara Falls will sail;
> And then no stage will be required
> To carry up the mail.
>
> —Niagara Falls Table Rock Album[1]

In the summer of 1841, a momentous event in the history of western navigation took place in the harbor of Oswego, New York, on Lake Ontario: the launching of the *Vandalia*, one of the first steam-driven screw-propelled ships in North America. First conceived in 1837, and assisted by the newly invented Ericsson propeller,[2] the 138-ton vessel was built by Alvin Bronson and friends of Oswego and intended solely for trade on the Welland Canal and the Lower Great Lakes (fig. 7.1). Covering an average of six miles per hour, and designed for heavy merchandise and luxury travel, the *Vandalia* was faster and more efficient than the old, clumsy side-wheeled steamers, still in use on the Great Lakes, that often caused damage to the walls of the canal. With much excitement did the *Oswego Whig* write that this new innovation in steam navigation "sheds a flood of glory upon the prospects of Oswego and the Welland Canal trade, absolutely blinding us with its bright radiance."[3] Across the lake in Upper Canada, the *Kingston Chronicle and Gazette* exclaimed with "great satisfaction" that they had "visited the new American vessel Vandalia" which, after making several stops at Kingston and Hamilton on Lake Ontario, provided evidence of "this great improvement in our commercial facilities." By *commercial facilities*, the newspaper

Figure 7.1. **The *Vandalia*.** One of the first steam-driven screw-propelled ships in North America. The *Vandalia* was built in Oswego, New York, specifically for trade on the Welland Canal. Courtesy of the Buffalo History Museum.

was referring not to a particular port, but an overall shared transportation circuit. It spoke to the north-south orientation of the Niagara–Great Lakes borderland resulting from the canal age.[4] The *Vandalia* was one of the most important cross-border developments in the history of North American transportation, heralded by the *St. Catharines Journal* as "a new era in the history of navigation."[5] As the *Vandalia* made her first trip through the Welland, "crowds of people turned out to see her, and a public dinner" was held in honor of the American proprietors at St. Catharines. In addition to its commercial and recreational potential, the *Vandalia*'s engines were said to have been constructed in the New York Auburn prison, indicating the continuing spirit of progress and reform that came out of the canal era in this region.[6]

Canadian-American cross-border cooperation and friendship continued to drive the transportation revolution even after the completion of the New

York and Welland Canal systems. During the period of enlargement and improvements to both transportation arteries, Canadian and American engineers, builders, workers, business leaders, and entrepreneurs actively looked across the border for jobs and related opportunities. American engineers Nathan Roberts, Benjamin Wright and James Geddes continued to provide much-praised service on the Welland Canal. In fact, problems relating to production, technology, and changes to the route after its ceremonial opening in 1829 might have undermined the Welland project were it not for their keen insight and experience. Well-known American engineer Alfred Barrett, who provided invaluable service in both New York and Upper Canada, was called to Montreal in 1837 where he was appointed resident engineer on the Lachine Canal. He later lost his life to the cholera pandemic while working on the Canadian transportation system. Meanwhile, American contractor Oliver Phelps maintained a permanent residence near the Welland Canal in Upper Canada but undertook a number of canal jobs with his son in New York State. These same American engineers and builders passed on their skills and expertise to local Upper Canadians like Thomas Coltrin Keefer, the son of the first president of the Welland Canal Company, who in 1838 apprenticed as an engineer on the Erie Canal before being appointed provisional engineer on the Welland Canal.[7] When Canada's leaders of transportation undertook additional improvements to the St. Lawrence River system, American engineers like Benjamin Wright were again called upon to conduct the surveys.

It was through such sustained and consistent Canadian-American connections that the entire northern trade corridor from the Upper Great Lakes to the ocean was made accessible to both countries. Indeed, it was owing in part to men like Benjamin Wright who suggested in an 1833 report that the private Welland Canal Company, then facing insolvency and in need of much rebuilding and repair, be purchased by the Canadian government and made a national work. Not long after Wright's findings, the fate of the private company was sealed when an 1837 Act of the Upper Canadian legislature provided for eventual government takeover.[8] The demise of the private Welland Canal Company just happened to coincide with the passing of one of its largest benefactors—John B Yates of Chittenango, New York. Before his untimely death in 1836, Yates regularly accompanied Canada's father of the Welland Canal, William Hamilton Merritt, on his visits to inspect the ongoing improvements to the St. Lawrence River system that both men regarded as crucial to the continuing commercial and recreational growth and prosperity of the borderland region. Later, a legislative report asserted that the Welland Canal Company would have collapsed if not for Yates's individual enterprise and resources, and it was to Yates that "the Country is mainly indebted for the early construction of this work."[9] A street bearing Yates's name in Merritt's hometown of St. Catharines today is testimony to this Yankee's unyielding support and friendship. Merritt himself acknowledged Yates as one "without whom the canal would never have been accomplished."[10]

If the United States supported and welcomed transportation developments in the northern province, improvements to the New York Canal system in the late 1830s were similarly praised by Upper Canadians. In 1835, the *Kingston Chronicle and Gazette* congratulated the United States on the enlargement program that was just getting under way on the Erie Canal. "We have much pleasure in copying the following interesting account of the proposed enlargement of this great work" commented the article, "the extraordinary successes attending the Erie Canal since its opening . . . must strike very forcibly, every friend of internal improvement."[11] Improvements to navigation, no matter on which side of the border, meant more efficient and cost-effective trade and transportation for inland merchants and shippers, while also bringing economic, commercial, agricultural, and recreational prosperity and improvement to the borderland region. Around the same time, it was anticipated that continuing improvements to the Welland Canal would result in "a very brisk trade" with Buffalo. Canadian newspapers similarly continued to announce employment opportunities on Oswego's ongoing public works projects.[12]

Improvements to transportation in the 1830s and 1840s brought opportunities to both sides of the border. The advantages that canals produced commercially and domestically for both countries, wrote the American *Merchants Magazine and Commercial Review* a few years later, "were important and striking." The same magazine continued that with "the aid of some short lines of canals formed to overcome the natural obstacles presented to navigation by the Falls of Niagara and the St. Lawrence, these great lakes are converted into a continuous line of water communication, penetrating upwards of two thousand miles into the remotest regions of North America, and affording an outlet for the produce of a large portion of this continent."[13] The enlargement of the Welland system was also of great advantage to Oswego as such improvements allowed for the passage of larger lake and steam vessels then sailing the Great Lakes. The *New York Commercial Advertiser* similarly applauded improvements on the Canadian system noting: "Our Canada friends are making great efforts to accomplish . . . the widening and deepening" of the Welland Canal. A Canadian correspondent, after witnessing developments on the improved Welland line, wrote that no one will hesitate to affirm "that William Hamilton Merritt is the Clinton of Canada."[14]

While the transportation revolution did not benefit everyone, the porous border did offer the promise of work to the hundreds of lesser-known individuals who sought economic opportunities in the neighboring countries. From the domestic servants and farmers who early left Upper Canada for higher wages on the New York canal system, to the thousands of unskilled laborers, subcontractors, and mechanics who found employment on Canada's Welland Canal, canalers moved freely back and forth across the border wherever jobs took them. Indeed, by the late 1830s when canal work virtually shut down in the United States because of the Panic, American canal workers, many accompanied by their families, moved to Upper Canada where construction jobs were to be had in the neighboring province.[15] However, the North American canal system also brought perils of disease and hardships that left too many workers with little in the way of prosperity. Others saw their land destroyed or, as in the case of some of Canada's aboriginal peoples, settled near the Welland Canal, a way of life forever disrupted by the transportation revolution. Still, the efforts of canal leaders and reformers to bring improvement to the poor lot of canal and boatmen who labored on the international waterways at least brought public attention to the problems that invariably followed on the heels of transportation development. Men like Alfred Barrett, William Hamilton Merritt, Gerrit Smith, Oliver Phelps, and many others, in association with the numerous American and Canadian missionary and tract societies, encouraged temperance and reform in both countries, providing yet another example of how the interlocking, interconnected canal network encouraged progress and good neighborhood across the northern border. The joint efforts of Canadian and American reformers also remind us that the religious fervor that swept through western New York's *burned over district* during the canal age was part of a larger borderland phenomenon.

Following the opening of the New York and Canadian canal systems, many of the continent's leading engineers and entrepreneurs turned their attention to railroad building, signifying yet another step forward in the ongoing North American transportation revolution. At the forefront of the iron rail was no other than Welland Canal founder William Hamilton Merritt who, while enjoying a picnic above Niagara Falls with his American wife, first conceived the idea of a railroad bridge spanning the Niagara

River. The growing profitability of the canal age, the rapid settlement of the Great Lakes region, and the potential to attract tourism and commerce to the region prompted Merritt once again to look to the United States for support in building a bridge that would facilitate the more rapid shipment of goods and passengers across the border. As finally realized in 1848, the internationally acclaimed Niagara Suspension Bridge drew thousands of visitors to the region, fulfilling Merritt's vision of promoting trade, tourism, and development with the neighboring Republic. Meanwhile, Great Britain's repeal of the Corn Laws during the same decade offered another incentive to unite both countries' economies as they adjusted to the larger economic and political changes taking place in faraway Europe. Though Canada's mercantile elite, especially in Montreal, initially panicked over the loss of protected markets in the mother country, borderland merchants and businessmen in the Great Lakes and farther west readily adapted by pushing for a free trade agreement between the United States and Canada. Led by William Hamilton Merritt, in association with America's leading abolitionist and friend of internal improvements Gerrit Smith, and Oswego's mercantile titan Alvin Bronson, the historic 1854 Reciprocity Treaty allowed American merchants and shipper's free and open access to the St. Lawrence River, while eliminating duties on agricultural goods across the United States–Canada border. Having spent much of their lives promoting transportation and communications across regions and boundaries, all three men valued and understood the importance of close economic and social ties between the neighboring countries.[16]

In recent years, the potential of canals and other internal improvements to revitalize the Canada–United States borderland economy has led to a renewed interest in rejuvenating canal systems for commerce, tourism, education, and recreation in both countries. A vital commercial, recreational, and cultural artery in its day, the New York canal system, almost two centuries later, is springing back to life. The Oswego and connecting Erie Canal have witnessed an appreciable jump in commercial traffic as a result of the Canadian grain industry increasingly turning its eye south and shipping goods through the New York canal system.[17] Millions of dollars to date have also been spent on the Erie Canal's National Heritage Corridor in an effort to bring commercial and recreational life back to this historic waterway. At the same time, Canada's creation of the Fort Erie Friendship Trail from Fort Erie to the Welland Canal with several shorter trails between the Peace Bridge and Buffalo likewise attests to the interconnectedness of the Canada–United States borderland and its potential to act as a stimulus to recreational growth in this region. Meanwhile, the more than one hundred thousand spectators who came from all over the world to witness American Nik Wallenda's historic tightrope walk over the Canadian Horseshoe Falls in 2012 affirmed Niagara's commercial and recreational importance to the larger borderland community, and a renewed interest in the role of tourism and spectacle in going beyond spotlighting Niagara Falls as a natural wonder of the world.

Reflecting back many years after the completion of the Welland Canal, American engineer William Gooding wrote a nostalgic letter to Canada's father of transportation, William Hamilton Merritt, about his years of service on the canal during the 1820s and 1830s, and how it endeared him to the neighboring country: "Having commenced my apprenticeship as an engineer on the Welland Canal," Gooding wrote, "it would be very natural to conclude that I had kept myself posted in relation to it ever since. . . . Perhaps it is because in the early stages of its development I was connected with it, and have always since felt a deep interest in its success." In the same letter, Gooding spoke about the unique Canada–United States relationship along the northern borderland, noting that American merchants in the far west were eager to see improvements to canals in both countries and in terms of exporting their products to the east they did not "give a straw" whether they were shipped on "the Canada or the American side."[18] Gooding simply affirmed what Merritt and his cohorts in both the United States and Canada already knew—the border presented a host of opportunities for Canadian-American merchants, businessmen, forwarders, and shippers on the Great Lakes who cared little about which flag their produce sailed under.

By analyzing the canal age as a transnational phenomenon, this book has offered an alternative view to nation-centered history which argues that Canadian and American canal planners deliberately followed an east-west orientation of trade and transportation that would enrich their individual countries. In contrast, I show how Canadians and Americans mutually promoted north-south transportation and communication lines in the Niagara–Great Lakes Basin that fostered cooperation and friendly competition between the neighboring peoples. Isolated from their respective metropolises, Canadians and Americans between the early 1790s and the late 1830s looked across the border for trade and cultural and recreational exchange. The difficulty and expense of getting goods to market, the need to overcome barriers to transportation and trade, a shared geography and history, and the establishment of a significant cross-border trade all demanded ambitious improvements on a previously unknown scale. Meanwhile, shared experiences and common responses to the War of 1812, and the desire to quickly reestablish personal and business ties across the international border, helped overcome any lingering enmity stemming from the conflict, while at the same time checking expansionist tendencies emanating from the United States' distant capital. In the postwar era, transnational linkages were developed more tangibly as both countries mutually undertook the construction of a vast, integrated, interconnected navigation system that encouraged westward development, market expansion, tourism, and improvement generally. As Gooding's letter confirmed, canals and other internal improvements had succeeded in overcoming both artificial and natural barriers in the Niagara–Great Lakes Basin, thereby transforming this once historically contested region into a porous and open cross-border community with lasting social, economic, and cultural connections. By forging transnational linkages and connections across their shared boundary, New York and Upper Canadian canal leaders and visionaries redefined the meaning and outcome of the North American canal age.

Notes

Chapter 1

1. Captain Basil Hall, *Travels in North America in the Years 1827 and 1828*, 2 vols. (Edinburgh: Printed for Cadell and Company, 1829) 1:175–76.

2. Hall, *Travels in North America*, 1:220.

3. The story of the competing routes is expanded on in chap. 3. See also Bukowczyk et al., *Permeable Border: The Great Lakes Basin as Transnational Region, 1650–1990* (Pittsburgh, PA: University of Pittsburgh Press, 2005), 30–31.

4. Canada's canal system was said to deliberately follow Laurentian lines (a system of transportation and trade that followed an east-west orientation) and was intended to challenge the Erie Canal. For a review of this scholarship see Roberta Styran and Robert Taylor, *The Great Swivel Link: Canada's Welland Canal* (Toronto, Ont.: The Champlain Society, 2001), xxviii–xxxi.

5. For more on the canal historiography in the United States see ch. 3, n.2.

6. John N. Jackson, *The Mighty Niagara: One River—Two Frontiers* (Amherst, NY: Prometheus Books, 2003), 277; Patrick McGreevy, *Stairway to Empire: Lockport, the Erie Canal, and the Shaping of America* (Albany, NY: SUNY Press, 2009), 19; Bruce Fisher, *Borderland: Essays From the US-Canadian Divide* (Albany, NY: SUNY Press, 2012), 5; *The Welland Canal Section of the St. Lawrence Seaway*, 1. http://www.greatlakes-seaway.com/en/pdf/welland.pdf.

7. Hugh Aitken, *The Welland Canal Company* (1954; repr., St. Catharines, Ontario: The Canadian Canal Society, 1997), 20. Originally published in 1954 by Harvard University; McGreevy, *Stairway to Empire*, 19.

8. Roberta Styran and Robert Taylor, *This Great National Object: Building the Nineteenth Century Welland Canals* (Montreal & Kingston: McGill-Queen's University Press, 2012), 19.

9. Aitken, *The Welland Canal Company*, 2.

10. This theme is expanded upon in chap. 6, esp. n.1. Elizabeth McKinsey, *Niagara Falls: Icon of the American Sublime* (Cambridge, UK: Cambridge University Press, 1985), 100; Patrick McGreevy, "The End of America: The Beginning of Canada," *The Canadian Geographer* 32, no. 4 (1988): 310.

11. Jane Errington, *The Lion, the Eagle and Upper Canada: A Developing Colonial Ideology* (Montreal: McGill-Queens University Press, 1987), 5 and 37; George Sheppard, *Plunder, Profit and Paroles: A Social History of the War of 1812 in Upper Canada* (Montreal & Kingston: McGill-Queen's University Press, 1994), 19–20; D. W. Meinig, *The Shaping of America: A Geographical Perspective on 500 Years of History*, 4 vols. (New Haven, CT: Yale University Press, 1993) 2:44.

12. Meinig, *The Shaping of America*, 2:41. Lauren McKinsey and Victor Konrad, "Borderlands Reflections: The United States and Canada" (Orono, ME: Borderlands Project, 1989), iii.

13. George Rogers Taylor for example, early stated that "the Erie Canal provided the spark which set-off a nation-wide craze for canal building," in G. R. Taylor, *The Transportation Revolution: 1815–1860* (New York: M. E. Sharpe, Inc., 1951), 34.

14. The term "lateral canal" diminishes the importance of the Oswego Canal as a viable trade artery *in its own right*, particularly when it is considered in connection with the Welland Canal. For references to the Oswego Canal as a lateral component of the larger Erie system see F. Daniel Larkin, *New York State Canals: A Short History* (New York: Purple Mountain Press, 1998), 42; Ronald Shaw, *Canals for a Nation: The Canal Era in the United States, 1790–1860* (Lexington, KY: The University of Kentucky Press, 1990), 45.

15. Aitken, *The Welland Canal Company*, 49; Thomas McIlwraith, "Transport in the Borderlands," in Robert Lecker, ed., *Borderlands: Essays in Canadian-American Relations*, (Toronto, Ont.: ECW Press, 1991), 62; F. Larkin, *New York Canals*, 50.

16. *Free Oswego Press*, Aug. 11, 1830.

17. Canal scholars frequently point out that in the long run, tonnage on the New York canal system far surpassed that on the Welland, and that the Oswego Canal served the interests of the United States far more than it did Canada. However, as will be discussed in greater length in this book, the Oswego-Welland line boosted market development, expansion, commercial and recreational traffic inland, and served the interests of local merchants, forwarders, and shippers on both sides of the border. For a comparison of the two routes see Donald Creighton, *The Empire of the St. Lawrence* (Toronto, Ont.: The Macmillan Company of Canada Limited, 1956), 251; Jackson, *The Mighty Niagara*, 130; Ronald Shaw, *Canals for a Nation*, 46–47; Bukowczyk et al., *Permeable Border*, 198, n.48; Thomas McIlwraith, "Freight Capacity and Utilization of the Erie and Great Lakes Canals before 1850," *Journal of Economic History* 36, no. 4 (December 1976) 853–75. This theme will be expanded upon in chap. 5.

18. Ronald Shaw notes for example that early improvements to the Oswego River were for the purpose of developing western New York, "to draw the trade of the west to the Hudson, and to divert that trade away from the outlet of Lake Ontario, which passed down the St. Lawrence River to Montreal." Shaw, *Canals For A Nation*, 30.

19. Lawrence H. Officer and Lawrence B. Smith, "The Canadian-American Reciprocity Treaty of 1855–1866," *The Journal of Economic History* 28, no. 4 (December 1968): 605; Thomas McIlwraith, "The Logistical Geography of the Great Lakes Grain Trade, 1820–1860" (PhD diss., University of Wisconsin, 1972), 91.

20. Douglas McCalla, *Planting the Province: The Economic History of Upper Canada, 1784–1870* (Toronto, Ont.: University of Toronto Press, 1993), 4–5; David M. Ellis, "The Rise of the Empire State, 1790–1820," *New York History* 56, no. 1 (1975): 14.

21. Officer and Smith, "The Canadian-American Reciprocity Treaty," 607. This point is also made in David Murray, "Criminal Boundaries: The Frontier and the Contours of Upper Canadian Justice, 1792–1840," *The American Review of Canadian Studies* 26, no. 3 (Autumn, 1996): 357–59.

22. Douglas McCalla, "QUETTON ST GEORGE, LAURENT," in *Dictionary of Canadian Biography*, vol. 6, University of Toronto/Université Laval, 2003– , accessed March 30, 2017, http://www.biographi.ca/en/bio/quetton_st_george_laurent_6E.html. Emily Cain, "Customs Collection—and Dutiable Goods: Lake Ontario Ports, 1801–1812," *Freshwater* 2, no. 2 (Autumn, 1987): 24.

23. David R. Moore, "Canada and the United States, 1815–1830" (PhD diss., University of Chicago, 1910), 121–22; Reginald Stuart, *United States Expansionism and British North America, 1775–1871* (Chapel Hill, NC: The University of North Carolina Press, 1988), 115.

24. McCalla, *Planting the Province*, 30; Neil McNall, *An Agricultural History of the Genesee Valley, 1790–1860* (Philadelphia, PA: University of Philadelphia Press, 1952), 102. Bukowczyk et al., *Permeable Border*, 52.

25. Few studies consider canals and reform as a North American phenomenon. More common are nation-centered studies that address the reform era during the Erie Canal's building and operation. Notable among the latter are Whitney R. Cross, *The Burned Over District: The Social and Intellectual History of Enthusiastic Religion in Western New York, 1800–1850* (New York: Harper and Row Publishers, 1950), 132 and 153–54. See also Roger Carp, "The Limits of Reform: Labor and Discipline on the Erie Canal," *Journal of the Early Republic* 10, no. 2 (Summer, 1990): 191–219; McGreevy, *Stairway to Empire*, 182–201; Shaw, *Erie Water West*, 219–235; and Carol Sheriff, *The Artificial River: The Erie Canal and the Paradox of Progress, 1817–1862* (Hill and Wang, 1996), 138–77. In Upper Canada, issues of temperance, morality and religious observance on the Welland Canal have received little attention with the exception of Styran and Taylor who analyze the changing nature of canal work on all four Welland Canals during the nineteenth and early twentieth centuries. See Styran and Taylor, *This Great National Object*, 289–302.

26. While this study is interested in the canal experience on both sides of the border, William Irwin writes that travelers to Niagara found the Erie Canal "to be a thrilling tourist attraction in and of itself." Irwin, *The New Niagara: Tourism,*

Technology and the Landscape of Niagara Falls, 1776–1917 (University Park, PA: The Pennsylvania State University Press, 1996), 14. See also chap. 6, n.2.

27. Although my purpose in this study is not a broad social history of the region or period, insofar as aspects of that social history speak to the story being documented, I do acknowledge women, Native peoples, and slaves whose stories are revealed through reformers and local canal leaders who supported their efforts.

28. As discussed in chap. 3, the question of the American threat to Canadian economic and cultural independence played itself out in the contest over the route of the New York canal system, and in the subsequent building of the Welland Canal as a strategic link in the Great Lakes–St. Lawrence water system.

29. For the conventional account of the New York canal system see McGreevy, *Stairway to Empire*, 154–55; Thomas F. McIlwraith, "Transport in the Borderlands, 1763–1920," in Lecker, ed., *Borderlands*, 61–62; Styran and Taylor, *This Great National Object*, 42; Shaw, *Erie Water West*, 403. See also chap. 3, n.2.

30. Meinig, *The Shaping of America*, 2:198.

31. Bukowczyk et al., *Permeable Border*, 34.

32. Bukowczyk et al., *Permeable Border*, 177.

Chapter 2

1. "Selections for the Emerald," *The Emerald (1810–1811)* 1, no. 15 (February 9, 1811): 172.

2. Late Loyalists were American-born settlers in Upper Canada who came during the 1790s not because of persecution as was the fate of the loyalist during the American Revolution, but for personal opportunity and advancement. Americans, particularly from New York and Pennsylvania, were attracted by offers of free or cheap land and other incentives like proximity to friends and family who remained in the United States. Alan Taylor, *The Civil War of 1812* (New York: Vintage Books, 2011), 8; Fred Landon, *Western Ontario and the American Frontier* (Toronto, Ont.: McClelland and Stewart Limited, 1967), 19. Douglas McCalla, *Planting the Province*, 15–16; Robert Bothwell, *Your Country, My Country: A Unified History of the United States and Canada* (Oxford, UK: Oxford University Press, 2015), 72.

3. Buckowczyk et al., *Permeable Border*, 26; Meinig, *The Shaping of America*, 2:44.

4. Robert Bothwell notes that even the original loyalists who came to Upper Canada helped forge cross-border ties in the Niagara–Great Lakes region. Speaking of the Loyalists who chose to go to Canada after the Treaty of Paris (1783) was signed, Bothwell observes: "Contact with the United States did not cease. Families exchanged letters and even visits, not just among the prominent, rich, and well connected, but on a much humbler level." Bothwell, *Your Country, My Country*, 69.

5. Robert L. Jones, *The History of Agriculture in Ontario, 1613–1880* (Toronto, Ont.: University of Toronto Press, 1946), 19. Alan Taylor *The Civil War of 1812*, 8 and 49–54.

6. Bothwell, *Your Country, My Country*, 72.

7. Errington, *The Lion, The Eagle, and Upper Canada*, 15 and 36. Meinig, *The Shaping of America*, 2:44; John Bartlett Brebner, *North Atlantic Triangle: The Interplay of Canada, the United States and Great Britain* (New Haven, CT: Yale University Press, 1945), 66.

8. Carl Benn, "Iroquois Warfare, 1812–1814," in R. Arthur Bowler, ed., *War Along the Niagara: Essays on the War of 1812 and its Legacy* (Youngstown, NY: Old Fort Niagara Association, Inc., 1991), 61. For the reference to native lands and the Welland Canal see Isabel T. Kelsay, "TEKARIHOGEN (1794–1832)," in *Dictionary of Canadian Biography*, vol. 6, University of Toronto/Université Laval, 2003, accessed March 31, 2017. http://www.biographi.ca/en/bio/ tekarihogen_1794_1832_6E.html.

9. Robert Hamilton to Peter Russell, November 1, 1789. Peter Russell Papers, Ontario Public Archives.

10. Bruce Wilson, "Patronage and Power: The Early Political Culture of the Niagara Peninsula," essay in Richard Merritt, et al., *The Capital Years: Niagara on the Lake, 1792–1796* (Toronto, Ont.: Dundurn Press, 1991), 59.

11. C. E. Cartwright, *Life and Letters of the Late Honourable Richard Cartwright* (Toronto, Ont.: Belford Brothers, 1876), 86.

12. Stuart, *United States Expansionism*, 33–34.

13. Marcus Lee Hansen, *The Mingling of the Canadian and American Peoples*, 2 vols. (New Haven, CT: Yale University Press, 1940) 1:65. Any suspicion or hatred that did exist toward the United States, writes another Canadian scholar was restricted to Upper Canada's small ruling elite in the distant capital at York. See Sheppard, *Plunder, Profit, and Paroles*, 9.

14. See ch. 6 for more on the Northern Tour and tourism at Niagara.

15. Kalm's account can be read in Frank Severance, *Studies of the Niagara Frontier* (Buffalo, NY, 1911), 325.

16. Ralph Izard, *An Account of a Journey to Niagara, Montreal and Quebec in 1765* (New York: Printed by Wilson Osborn, 1846), 5–9.

17. Diary of Anne Powell on her voyage from Montreal to Detroit with her brother W. D. [William Dummer] Powell (later Chief Justice of Upper Canada). Jarvis family fonds, 1789–1847, n.d. RG 563, Brock University Archives. The diary has also been published in *The Magazine of American History*(July 1880): 37–47.

18. Creighton, *The Empire of the St. Lawrence*, 72. William Kingsford, *The History of Canada*, 10 vols. (Toronto, Ont.: Roswell & Hutchinson, 1887–1898) 7:28.

19. R. W. Riddell, *The Life of William Dummer Powell, first judge at Detroit and fifth chief justice of Upper Canada* (Lansing MI: Michigan Historical Commission, 1924), 61. Hereafter *Life of William Dummer Powell*.

20. See also Samuel Wilkenson, "The Life of the Keel Boatmen," *Buffalo and Erie County Historical Society Publications* (Buffalo, NY: 1902) 5:179–82. As will be discussed in chap. 5, during the transportation revolution a reform movement emerged in the Niagara–Great Lakes borderland region that sought to bring improvement to the hundreds of canal and boatmen who labored on the interconnected water system.

21. "ART. VI." *The Monthly Magazine, and American Review (1799–1800)* 1, no. 2 (May, 1799): 119. Richard Merritt's article "Early Inns and Taverns: Accommodation, Fellowship and Good Cheer" considers the connection between inns and transportation systems in eighteenth-century Niagara, Upper Canada, in Merritt, et al., *The Capital Years*, 187–222. Roads connecting Canada and the United States along the Niagara borderland before the War of 1812 were also eagerly promoted by merchants, land speculators, travelers, and government officials hoping to stimulate prosperity and settlement here. See William Wyckoff, *The Developer's Frontier: The Making of the Western New York Landscape* (New Haven, CT: Yale University Press, 1988), 58.

22. Wyckoff, *The Developer's Frontier*, 56.

23. McCalla, *Planting the Province*, 134; Wyckoff, *The Developer's Frontier*, 53 and 99; Landon, *Western Ontario and the American Frontier*, 129; Errington, *The Lion, The Eagle and Upper Canada*, 37.

24. "To Alexander Hamilton from Timothy Pickering, 3 January 1793," in Harold C. Syrett, ed., *The Papers of Alexander Hamilton*, 27 vols. (New York: Columbia University Press, 1961) 13: 449–50. See also McCalla, *Planting the Province*, 134, and Merritt, et al., *The Capital Years*, 203.

25. From the *Upper Canada Gazette*, September 20 and October 11, 1797.

26. *Buffalo Gazette*, February 5, 1812.

27. Called the cradles because it was actually built in three sections that carried goods in stages up the steep incline. Michael M. McConnell, *Army and Empire: British Soldiers on the American Frontier 1758–1775* (Lincoln, NE: University of Nebraska Press, 2004), 17.

28. Brian L. Dunnigan, "Portaging Niagara," *Inland Seas* 42, no. 3 (Fall 1986): 218.

29. George Seibel, *The Niagara Portage Road: 200 Years 1790–1990* (Niagara Falls, Ont.: The City of Niagara Falls Canada, 1990), 10.

30. *The Life of William Dummer Powell*, 66.

31. W. D. Howells, Mark Twain et al., *The Niagara Book* (New York: Doubleday, Page and Company, 1901), 101–02.

32. Seibel, *The Niagara Portage Road*, 12.

33. William W. Campbell, *The Life and Writings of De Witt Clinton* (New York: Baker and Scribner, 1849), 127.

34. Richard Switzer, *Chateaubriand's Travels in America* (Lexington: The University of Kentucky Press, 1969), 21.

35. *The Life of William Dummer Powell*, 66.

36. *Upper Canada Gazette*, July 11, 1793.

37. Future Erie Canal founder De Witt Clinton expands on this point in the *Letters on the Natural History and Internal Resources of the State of New York, by Hibernicus* [*De Witt Clinton*] (New York, 1822), 124.

38. John Maude, *Visit to the Falls of Niagara in 1800* (London: Longmans, Rees, Orme, Brown & Green, 1826), 132.

39. "Canadian Letters, Description of a Tour Thro' the Provinces of Lower and Upper Canada, 1792–1793," in Merritt, et al., *The Capital Years*, 190–91.

40. "T.C. A Ride to Niagara," in Charles M. Dow, *Anthology and Bibliography of Niagara Falls*, 2 vols. (Albany, NY: J. B. Lyon Company, Printers, 1921) 2:1188.

41. Campbell, *The Life and Writings of De Witt Clinton*, 125–26.

42. Quoted in Severance, *Studies of the Niagara Frontier*, 366.

43. *Niagara! The Eternal Circus* (Toronto, CAN: Doubleday Canada Limited, 1979), 84.

44. "Article 1—no Title," *The Literary Magazine, and American Register (1803–1807)* 2, no. 12 (September, 1804): 432. See also Irwin, *The New Niagara*, 3; McKinsey, *Niagara Falls*, 37.

45. For an account of the early history of the Black Rock ferry see Charles Norton, *The Old Black Rock Ferry* in the Buffalo and Erie County Historical Society Publications (Buffalo, NY: 1879) 1:91–112.

46. E. Cruikshank, *The Correspondence of Lieut. Governor John Graves Simcoe, with allied documents relating to his administration of the government of Upper Canada*. 5 vols. (Toronto, Ont.: 1923–1931) 4:352–53.

47. Dow, *Anthology and Bibliography of Niagara Falls*, 1:67–68. Discussing the commercialization of the Falls after the War of 1812, McKinsey writes that boatmen deliberately made the ferry crossings appear more hazardous than they were in order to attract more tourists. McKinsey, *Niagara Falls*, 135 and 150. The significance of the Niagara ferry as a popular internal improvement on the Northern Tour is also discussed in chap. 6.

48. Elizabeth Simcoe and J. Ross Robertson, *The Diary of Mrs. John Graves Simcoe* (Toronto, Ont.: William Briggs, 1911), 128.

49. "Monthly Register," *The New York Magazine, Or Literary Repository (1790–1797)* (November, 1797): 613.

50. Seibel, *The Niagara Portage Road*, 109.

51. Campbell, *The Life and Writings of De Witt Clinton*, 127.

52. J. P. Merritt, *Biography of the Honourable William Hamilton Merritt* (St. Catharines, Ont.: E. S. Leavenworth Est., 1875), 60. Hereafter cited, J. P. Merritt, *Biography*.

53. See *Upper Canada Gazette*, April 26, 1797.

54. *Upper Canada Gazette*, May 31, 1797.

55. John Melish, *Travels through the United States of America, in the years 1806 & 1807, and 1809, 1810, & 1811; including an account of passages betwixt America and Britain, and travels through various parts of Britain, Ireland, & Canada* (Philadelphia, PA: Printed for the Author; London, Reprinted for George Cowie and Co, 1818), 489. Online Text. Retrieved from the Library of Congress, https://www.loc.gov/item/01025002/ (accessed April 03, 2017).

56. See for example *Upper Canada Gazette*, February 8, 1797.

57. Brian Tobin and Elizabeth Hulse, eds., *The Upper Canada Gazette and its Printers, 1793–1849* (Toronto, Ont.: Ontario Legislative Library, 1993), 7. For more on newspapers along the Niagara borderland see Alan Taylor, *The Civil War of 1812*, 70–71.

58. The problem of smuggling in the Niagara–Great Lakes borderland during the canal age will be explored in later chapters.

59. notesonniagara1732niag, August 14, 1791.

60. The Duke de la Rochefoucauld-Liancourt, *Travels through the United States of North America . . . in the years 1795, 1796, and 1797* (London: R. B. Phillips, 1799), 270.

61. notesonniagara1732niag, August 14, 1791.

62. As quoted in Israel Ira Rubin, *New York State and the Long Embargo* (PhD diss., New York University, 1961), 129.

63. John Melish, *Travels in the United States of America, in the Years 1806 & 1807*, 539. This view is also substantiated by William Jennings, *The American Embargo 1807–1809: With particular reference to its effect on industry* (Iowa City, IA: The University, 1921), 113–14.

64. Douglas McCalla, "QUETTON ST GEORGE, LAURENT," in *Dictionary of Canadian Biography*, vol. 6, University of Toronto/Université Laval, 2003– , accessed March 30, 2017, http://www.biographi.ca/en/bio/quetton_st_george_laurent_6E.html.

65. Emily Cain, "Customs Collection—and Dutiable Goods," 24; Stuart, *United States Expansionism*, 51.

66. Walter W. Jennings, *The American Embargo 1807–1809*, 113–14.

67. Emily Cain, "Provisioning Lake Ontario Merchant Schooners, 1809–1812," *Fresh Water* 3, no. 1 (Summer 1988): 21. See also Alan Taylor, *The Civil War of 1812*, 119.

68. John Melish, *Travels Through the United States of America in the Years 1806 & 1807*, 539; *Letters on the Natural History and Internal Resources of the State of New York by Hibernicus [De Witt Clinton]*. (New York, 1822), 209.

69. This theme will be discussed in ch. 6.

70. O. Turner, *History of the Pioneer Settlement of Phelps and Gorham's purchase and Morris' Reserve: Embracing the counties of Monroe, Ontario, Livingston, Yates, Steuben, most of Wayne and Allegany, and parts of Orleans, Genesee, and Wyoming: to which is added, a supplement or extension of the pioneer history* (Rochester, NY: W. Alling,

1851), 106–07; *Illustrated Historical Atlas of Erie County, New York* (New York: F. W. Beers and Company, 1880), 6. See also Stuart, *United States Expansionism*, 31.

71. Neil Adams McNall, *An Agricultural History of the Genesee Valley, 1790–1860* (Philadelphia, PA: University of Philadelphia Press, 1952), 102. David Ellis also comments on the importance of local trades in "Rise of the Empire State, 1790–1820," 14.

72. Isaac Weld, *Travels Through the States of North America, and the Provinces of Upper and Lower Canada*. 2 vols. (London: Printed for John Stockdale, Piccadilly, 1799) 2: 90. Online Text. Retrieved from the Library of Congress, https://www.loc.gov/item/17013770/ (accessed April 3, 2017).

73. Wyckoff, *The Developer's Frontier*, 70–71 and 115.

74. The Duke de la Rochefoucauld-Liancourt, *Travels Through the United States of North America*, 147.

75. Stuart, *United States Expansionism*, 34.

76. Douglas McCalla notes that local trades are the least documented before 1850 and yet a large and diverse range of economic activities occurred not only within the province but across the border with the adjacent states. McCalla, *Planting the Province*, 4–10. Hugh Aitken also notes that "a brisk trade across the lakes" occurred after 1796. Aitken, *The Welland Canal Company*, 12.

77. O. Turner, *Pioneer History of the Holland Purchase of Western New York: Embracing some account of the ancient remains, and a history of pioneer settlement under the auspices of the Holland Company; including reminiscences of the War of 1812; the origin, progress and completion of the Erie Canal, etc., etc., etc.* (Buffalo, NY: Jewett, Thomas & Co., 1849), 310–12.

78. By the War of 1812, however, trade patterns seemed to have reversed themselves because Upper Canadians were now in the profitable market of raising cattle on their land with the view of exporting them to the United States. See Jones, *History of Agriculture in Ontario*, 32.

79. The Credit River is at the far-western end of Lake Ontario and was abundant with salmon. Travelers on the Northern Tour frequently commented on the large quantity of salmon and other varieties of fish in the New York and Lake Ontario water systems.

80. O. Turner, *History of the Pioneer Settlement*, 410–11. As noted by more than one contemporary account, New Yorkers and Vermonters were "famous for making butter and cheese." See Jones, *History of Agriculture in Ontario*, 31.

81. R. W. Dunfield, *The Atlantic Salmon in the History of North America* (Ottawa, Can.: Department of Fisheries and Oceans, 1985), 74–75.

82. E. B. O'Callaghan, *The Documentary History of the State of New York*, 3 vols. (Albany, NY: Weed, Parsons & Co., Public Printers, 1849), 2:1164.

83. O'Callaghan, *The Documentary History of the State of New York*, 2:1149.

84. *Upper Canada Gazette*, October 4, 1797.

85. *Upper Canada Gazette*, December 30, 1797.

86. Orasmus H. Marshall, *The Niagara Frontier: Embracing Sketches of its Early History, and India, French, and English Local Names* (Buffalo, NY, 1865); W. Bird, *Early Transportation, New York State*. Buffalo Historical Society *Publications* (Buffalo, NY, 1880), 2:19.

87. La Duke de la Rochefoucauld-Liancourt, *Travels Through the United States of North America*, 153.

88. Wyckoff, *The Developer's Frontier*, 99.

89. H. Perry Smith, *History of the City of Buffalo and Erie County*, 2 vols. (New York: D. Mason & Company, Publishers, 1884) 1:81.

90. Hansen, *The Mingling*, 1:76.

91. McCalla, *Planting the Province*, 30 and 24.

92. Customs houses were established on the American side of the border in 1799 and on the Canadian side in 1802. Errington, *The Lion, the Eagle and Upper Canada*, 37. See also Seibel, *The Niagara Portage Road*, 140.

93. D. B. Read, *The Life and Times of Gen. John Graves Simcoe* (Toronto, Ont.: George Virtue, Publisher, 1890), 271; Aitken, *The Welland Canal Company*, 12.

94. Hamilton was discussing the importance of the fur trade to New York and the nation but his broader views on the province can be read in "The Defence No. XII," New York, September 2–3, 1795, and New York, April 22, 1796, in Syrett ed., *Papers of Alexander Hamilton*, 13:217–31. Reginald Stuart discusses Hamilton's attitudes toward the provinces in Stuart, *United States Expansionism*, 28–30; and Reginald Stuart, "Special Interests and National Authority in Foreign Policy: American-British Provincial Links during the Embargo and War of 1812," *Diplomatic History* 8, no. 4 (1984): 312.

95. "Monthly Register," *The New York Magazine, Or Literary Repository (1790–1797)* 4, no. 7 (July, 1793): 446.

96. "Original Poetry," *The New York Magazine, Or Literary Repository (1790–1797)* 6, no. 6 (June, 1795): 375.

97. Creighton, *The Empire of the St. Lawrence*, 135–36.

98. John Bartlett Brebner, *North Atlantic Triangle*, 80.

99. Russell's speech can be read in the *Upper Canada Gazette*, June 14, 1797.

100. John C. Ogden, *A Tour Through Upper and Lower Canada. By a Citizen of the United States* (Litchfield, 1799) in Dow, *Anthology and Bibliography of Niagara Falls*, 2:1185.

101. Howells, Mark Twain et al., *The Niagara Book*, 99.

102. Louis Hennepin, *A New Discovery of a Vast Country in America* (London: Printed for Henry Bonwicke, 1699), 23 and 215.

103. Frank Severance, *An Old Frontier of France*. 2 vols. (New York: Dodd, Mead and Company, 1917) 1: 198–210. See also Peter A. Porter, "The Niagara Ship Canal," unpublished papers and misc. clippings, The Buffalo History Museum.

104. Irwin, *The New Niagara*, 8.

105. "Back Material 1—no Title," *The Philadelphia Minerva, Containing a Variety of Fugitive Pieces in Prose, and Poetry, Original and Selected (1795–1798)* 1, no. 27 (August 8, 1795): 1.

106. McKinsey, *Niagara Falls*, 127.

107. Todd Shallat, *Structures in the Stream: Water, Science and the Rise of the United States Army Corp of Engineers* (Austin, TX: University of Texas Press, 1994), 58.

108. Melvin Hernandon, "A Grandiose Scheme to Navigate and Harness Niagara Falls," *NYHSQ* 58 (Jan 1974): 7. See also Irwin, *The New Niagara*, 9.

109. William Tatham, *The Political Economy of Inland Navigation* (London: Robert Faulder, 1799), 103.

110. David Hosack, *Memoir of De Witt Clinton* (New York: J. Seymour, 1829), 92–93 and 284.

111. Noble E. Whitford, *History of the Canal System of the State of New York*, 2 vols. (Albany, NY: Brandow Printing Company, 1905) 1:23–25; George Geddes, "The Erie Canal," Buffalo and Erie County Historical Society Publications (Buffalo, NY:1880) 2:267; O. Turner, *Pioneer History of the Holland Purchase of Western New York*, 619.

112. Hansen, *The Mingling*, 1:11.

113. "Report of Mr. Weston to the Directors of the Western and Northern Inland Lock Navigation Companies," Albany, December 23, 1795, in *American State Papers*, Documents, Legislative and Executive, of the Congress of the United States, . . . March 3, 1789, and ending March 3, 1809 (Washington, DC: Gales and Seaton, 1834), 1, class X, Misc.: 775.

114. Philip Lord, Jr., *The Navigators: A Journal of Passage on the Inland Waterways of New York, 1793* (Albany, NY: New York State Museum, 2003), 1.

115. Larkin, *New York State Canals*, 14; Shaw, *Erie Water West*, 19.

116. See Philip Lord, Jr., "The Mohawk/Oneida Corridor: The Geography of Inland Navigation," in David Curtis Scaggs, ed., *The Sixty Years War for the Great Lakes* (Lansing, MI: Michigan State University Press, 2010).

117. Robert Troup, *A Letter to the Hon. Brockholst Livingston, Esq., on the Lake Canal Policy of the State of New York* (Albany, NY, 1822), 4.

118. Irwin, *The New Niagara*, 13.

119. Shaw, *Erie Water West*, 17.

120. The competition of the new canals encouraged lower prices on freight wagons. John Melish, *Travels Through the United States of America in the Years 1806 and 1807*, 546.

121. "An Itinerary to Niagara Falls in 1809," in *The Pennsylvania Magazine of History and Biography*, XXIV (1900): 200–02.

122. Elkanah Watson, *History of the Rise, Progress, and Existing Condition of the Western Canals of the State of New York* (Albany, NY: Packard and Van Benthuysen Printers, 1820), 100.

123. Winslow C. Watson, ed., *Men and Times of the Revolution, or Memoirs of Elkanah Watson, including Journals of Travels in Europe and America from 1777–1842* (New York: Dana and Company Publishers, 1856), 321.

124. Anticipator. From the Albany Register, *The Time Piece; and Literary Companion (1797–1798)* 1, no. 7 (March 27, 1797): 26.

125. Watson, *History of the Rise, Progress, and Existing Condition of the Western Canals in the State of New York*, 100. As discussed in chap. 3, the Holland Land Company also took a keen interest in the Niagara Canal in 1798. See chap. 3, n.19.

126. There are no book-length studies on the Niagara Canal but Noble Whitford devotes a chapter to this work. Whitford, *History of the Canal System of the State of New York*, 1: chap. 7,

127. As discussed in later chapters, a canal around Niagara Falls, whether envisioned on the American or Canadian side, has conventionally been viewed as part of the larger commercial and military contest between Great Britain and the United States.

128. In addition, Williamson early pushed for a regularized ferry service across the Niagara River at Black Rock to stimulate cross-border trade. Helen I. Cowan, *Charles Williamson: Genesee Promoter: Friend of Anglo-American Rapproachment* (First Edition printed by the Rochester Historical Society, 1941. Reprinted in 1973 by Augustus M. Kelley Publishers, Clifton, NJ), 46–52 and 116 and 164–65. See also Wyckoff, *The Developer's Frontier*, 97.

129. *Laws of the State of New York Passed at the Sessions of the Legislature Held in the Years 1797, 1798, 1799, and 1800, inclusive* (Albany, NY: Weed, Parson, and Company Printers, 1887), 21st session, 5th of April, 1798, 271.

130. Robert Troup, *A Letter to the Hon. Brockholst Livingston*, 5–6. See also Noble Whitford, *History of the Canal System of the State of New York*, 1:45–46.

131. Anticipator. From the *Albany Register, The Time Piece; and Literary Companion (1797–1798)* 1, no. 7 (March 27, 1797): 26.

132. *Laws of the State of New York Passed at the Sessions of the Legislature Held in the Years 1797, 1798, 1799, and 1800*, 271.

133. Robert Troup, *A Letter to the Hon. Brockholst Livingston*, 6.

134. Peter Way, *Common Labour: Workers and the Digging of North American Canals, 1780–1860* (Cambridge, UK: Cambridge University Press, 1993), 22–23.

135. Shaw, *Canals For A Nation*, 19.

136. The conflict between progress and prosperity during the nineteenth-century canal age will be developed in later chapters.

137. Maude, *Visit to the Falls of Niagara in 1800*, 164–65.

138. John L. Larson, *Internal Improvement: National Public Works and the Promise of Popular Government in the Early United States* (Chapel Hill, NC: The University of North Carolina Press, 2001), 29.

139. *Bill to Improve and Amend the Communication Between Lakes Erie and Ontario, By Land and Water* (Niagara: S&G Tiffany, Printers to the Province, 1799).

140. Wilson, "Patronage and Power," in Merritt, et al., *The Capital Years*, 59.

141. Bruce Wilson, *The Enterprises of Robert Hamilton: A Study of Wealth and Influence in Early Upper Canada, 1776–1812* (Ottawa, Canada: Carleton University Press, 1983), 74 and 81.

142. Jones, *History of Agriculture in Ontario*, 26–27.

143. Seibel, *The Niagara Portage Road*, 31.

144. Wilson, "Patronage and Power," in Merritt et al., *The Capital Years*, 59.

145. For the reference to Hamilton and Ellicott see O. Turner, *Pioneer History of the Holland Purchase of Western New York*, 416–17. For Hamilton's relationship with Porteous see "Robert Hamilton, The Founder of Queenston," Buffalo and Erie County Historical Society *Publications* (Buffalo, NY, 1903) 6:92–94. Hamilton's relationship with Porteous lasted from 1789 to the latter's death in 1799.

146. First quote comes from Wilson, "Patronage and Power," in Merritt et al., *The Capital Years*, 59. For reference to Canadians purchasing American-made wagons see *Canada Constellation*, November 23, 1799.

147. G. P. de T. Glazebrook, *A History of Transportation in Canada* (Toronto, Ont.: The Ryerson Press, 1938), 67.

148. *Upper Canada Gazette*, October 26, 1796.

149. *Canada Constellation*, Niagara, December 7, 1799. See also Glazebrook, *A History of Transportation in Canada*, 67.

150. All aspects of Hamilton's commercial and portaging interests depended on open and friendly relations with the United States. Hamilton's success at Niagara was tied to the American market. Wilson, *The Enterprises of Robert Hamilton*, 103.

151. The rapids at Fort Erie also posed a barrier to traffic on the First Welland Canal. The latter problem, as discussed in chapt. 4, was overcome by utilizing the Black Rock ship lock on the American side of the river further attesting to the permeable border during the age of transportation.

152. *A Bill To Improve and Amend The Communication Between The Lakes Erie and Ontario, By Land and Water* (Niagara: Printed by S&G Tiffany, Printers to the Province, 1799). See also Wilson, *The Enterprises of Robert Hamilton*, 69 and 85.

153. Isaac Weld, *Travels through the states of North America, and the provinces of Upper and Lower Canada during the years 1795, 1796, and 1797.* 2:137.

154. New York State's role in the building of the Welland Canal is discussed in chap. 4.

Chapter 3

1. Dow, *Anthology and Bibliography of Niagara Falls*, 2:776.

2. Much writing on the canal age, both academic and popular, emphasizes nationalism and commercial rivalry between the two countries. Ronald Shaw set the tone of this scholarship when he argued that "rivalry with Canada for the trade of the old Northwest" and "the prospect of renewed war with Great Britain" influenced the location of the Erie Canal's route and the timing of construction. According to Shaw, if the Erie Canal had followed the Mohawk and Oswego River channel to Lake Ontario, with an additional canal cut around Niagara Falls, the western trade, having once reached Lake Ontario, would have been lost to the St. Lawrence River

and Montreal market. Instead, America's leaders deliberately built a canal to Lake Erie, avoiding the Lake route and the problem of Canadian competition altogether. Ronald Shaw, *Erie Water West*, 402–03. Similarly, Gerard Koeppel's engaging *Bond of Union* looks at the Erie Canal as a monumental achievement that provided the first great bond between the seaboard interests in the east and America's far west, while also stating that Upper Canada's Solicitor General early "lamented" the neighboring canal's success. George Koeppel, *Bond of Union*, 395. In his popular book, entitled *Wedding of the Waters*, Peter Bernstein argues that the Erie Canal prevented the breakup of the American empire, while stating that Canada took "a less cheerful view" of the American canal and when it became apparent that the New York channel was siphoning away so much traffic, Canadians feverishly began building the Welland Canal to offset the commercial losses. Bernstein, *Wedding of the Waters* (New York: W. W. Norton and Company, 2005), 352. The argument of Canadian-American competition and rivalry is also stated in Styran and Taylor, *This Great National Object*, 19. In 1956, William R. Willoughby also argued that work on the Erie Canal provided the major incentive for the construction of the Welland Canal that "would defeat the American grand design of diverting traffic from Canada to American channels . . . thus giving a new turn to the historic rivalry between the St. Lawrence waterway and the Mohawk-Hudson transportation route—a rivalry not yet finished." W. R. Willoughby, "The Impact of the Erie Canal," *Niagara Frontier* 3, no. 1 (Spring 1956): 38–47 and 42.

3. Landon, *Western Ontario and the American Frontier*, 1.

4. *Kingston Gazette*, February 12, 1811.

5. Wright is referred to in both academic and popular writings as the "American father of engineering," but given his invaluable contribution to both sides of the Canada/United States border during the canal age, it is perhaps more appropriate to refer to him as "North America's father of engineering."

6. *Laws of the State of New York in Relation to the Erie and Champlain Canals, together with the Annual Reports of the Canal Commissioners, and other Documents*, 2 vols. (Albany, NY, 1825) 1:13. Hereafter *Canal Laws*.

7. Philip Lord, Jr., *The Navigators: A Journal of Passage on the Inland Waterways of New York, 1793*, 1; "WATERWAYS WEST! Or How New Yorkers got across the state before the Erie Canal and major roadways." http://www.nysm.nysed.gov/press/waterways-west-or-how-new-yorkers-got-across-state-erie-canal-and-major-roadways.

8. The 1808 Act of the New York legislature requested that surveys be made for (1) a communication between Oneida and Lake Ontario, (2) the Niagara River, and (3) an interior route without descending to, or passing, through Lake Ontario. *Canal Laws*, 1:13.

9. See *Canadian Geographic Magazine*, Alex Hutchinson, "New Energy," April 2013 issue.

10. Simon DeWitt to Joseph Ellicott, August 24, 1808, in *The Holland Land Company—Buffalo–Black Rock Harbor Papers* (Buffalo, NY: Buffalo Historical Society, 1910), 16–17. Hereafter cited *Buffalo–Black-Rock Harbor Papers*.

11. Wyckoff, *The Developer's Frontier*, 99; Wilson, *The Enterprises of Robert Hamilton*, 74; Jackson, *The Mighty Niagara*, 107; Cain, "Customs Collection—and Dutiable Goods," 22. Marvin Rapp, "New York's Trade on the Great Lakes, 1800–1840," *New York History* 39, no. 1 (January 1958): 32, n.22.

12. Perry Smith, *History of the City of Buffalo*, 1:81. For more information on the early development of the borderland economy see Wyckoff, *The Developer's Frontier*, 99; Errington, *The Lion, the Eagle and Upper* Canada, 37; and McCalla, *Planting the Province*, 24.

13. Robert Bingham edition, *Reports of Joseph Ellicott*, 2 vols. (Buffalo, NY: Buffalo Historical Society, 1937–41) 1:144–45.

14. See chap. 2, n.145.

15. Wyckoff, *The Developer's Frontier*, 74 and 94 and 99; Jackson, *The Mighty Niagara*, 106–07.

16. John Horton, *History of Northwestern New York: Erie, Niagara, Wyoming, Genesee and Orleans Counties*, 3 vols. (New York: Lewis Historical Publishing, Inc., 1947) 1:30; Alan Taylor, *The Civil War of 1812*, 117.

17. Charles E. Brooks, *Frontier Settlement and Market Revolution: The Holland Land Purchase* (Ithaca, NY: Cornell University, 1996), 37; Wyckoff, *The Developer's Frontier*, 100; Horton, *History of Northwestern New York*, 1:29–30.

18. Joseph Ellicott to Paul Busti, June 9, 1810. Quoted in Shaw, *Erie Water West*, 43.

19. Following the 1798 Act of the New York Legislature incorporating the Niagara Canal Company, the Holland Company employed an "engineer and mineralogist" to investigate "the expenses and wisdom" of the project. Robert Bingham, *Reports of Joseph Ellicott*, 1:28–32.

20. *Buffalo–Black Rock Harbor Papers*, 17–19; Alfonso M. Ressa, *Paolo Busti: Chapter of American History, 1798–1824* (Philadelphia, PA: [s.n.], 1957), 16.

21. Paul Busti to Joseph Ellicott, May 17, 1811, in *Buffalo–Black-Rock Harbor Papers*, 19.

22. T.C., *A Ride to Niagara in 1809* (Rochester, NY, 1915), 17 and 32. Cooper's observations can also be read in Koeppel, *Bond of Union*, 64.

23. John Melish, *Travels Through the United States of America in the Years 1806, 1807*, 508–09.

24. *Canal Laws*, 1:46. The legislative resolution is also in the Peter Porter Papers entitled *State of New York, in Senate*, March 13, 1810, Roll 11, Misc. The Buffalo History Museum.

25. John Melish, *Travels Through the United States of America in the Years 1806, 1807*, 513.

26. *American State Papers. Documents, Legislative and Executive, of the Congress of the United States, . . . March 3, 1789, and ending March 3, 1809* (Washington, DC: Gales and Seaton, 1834), 1, class X, Misc.: 734.

27. Ibid, 776.

28. For a lively account of the 1810 canal expedition see Bernstein, *Wedding of the Waters*, 308–21.

29. Taken from an earlier report in the *Quebec Gazette* and reprinted in the *Kingston Gazette*, May 7, 1811.

30. Campbell, *Life and Writings of De Witt* Clinton, 32.

31. Campbell, *Life and Writings of De Witt Clinton*, 45.

32. Campbell, *Life and Writings of De Witt Clinton*, 47–48 and 54. The Western Company's progress in this region can be read in Lord, Jr., *The Navigators*, 21–28.

33. Others, too, were struck by both the utility and beauty of improvements in this region. One tourist wrote at Little Falls that the area "was beautiful to the extreme . . . there are several locks here by which boats proceed round the Falls. All very interesting and worth seeing." See "A Niagara Falls Tourist of the Year 1817 Being the Journal of Captain Richard Langslow," Buffalo and Erie County Historical Society Publications (Buffalo, NY, 1902) 5:113. Koeppel writes that Little Falls became a major attraction on the Erie Canal, in *Bond of Union*, 308.

34. Campbell, *Life and Writings of De Witt Clinton*, 62.

35. Few if any accounts mention the Commissioner's positive reactions to Canada and the Canadian market during the 1810 tour. In general, competition and commercial rivalry with the neighboring country is emphasized or alternatively, the commissioners' positive reflections on Canada are not mentioned at all. See, for example, Archer B. Hulbert, *Historic Highways of America* (Cleveland, OH: The A. H. Clark Company, 1902), 14:54; Koeppel, *Bond of Union*, 77–85; Bernstein, *Wedding of the Waters*, 145–51.

36. Campbell, *Life and Writings of De Witt Clinton*, 51 and 72.

37. Campbell, *Life and Writings of De Witt Clinton*, 77. See also Cain, "Customs Collection—and Dutiable Goods," 22. As will be discussed in a later chapter, travelers on the Northern Tour regularly commented on the salt trade, noting its destination to Canada and the west.

38. "A Ride to Niagara," *The Port-Folio (1801–1827)* 4, no. 3 (September, 1810): 230. See also Campbell, *Life and Writings of De Witt Clinton*, 112–13.

39. Campbell, *Life and Writings of De Witt Clinton*, 82.

40. Campbell, *Life and Writings of De Witt Clinton*, 74 and 85; For a reference to the Geddes map of the Oswego Canal in 1809 see *Canal Laws*, 1: 33.

41. David M. Ellis, "Rise of the Empire State, 1790–1820," 14; McCalla, *Planting the Province*, 24.

42. For more on the embargo and the profitable trade see Brooks, *Frontier Settlement and Market Revolution*, 37; Errington, *The Lion, the Eagle and Upper Canada*, 37; Stuart, *United States Expansionism*, 51; Wyckoff, *Developer's Frontier*, 100.

43. McNall, *Agricultural History*, 99.

44. Campbell, *Life and Writings of De Witt Clinton*, 82.

45. Cain, "Customs Collection—and Dutiable Goods," 24.

46. J. C. A. Stagg, *Mr. Madison's War: Politics, Diplomacy and Warfare in the Early American Republic, 1783–1830* (Princeton, NJ: Princeton University Press, 1983), 41. David Ellis also notes that "the inhabitants of the Champlain Valley, Oswego area, and Genesee country challenged the restriction" imposed by the embargo. "Moreover, the government did not enforce the law as strictly in trading with Canada." Ellis, "Rise of the Empire State," 12.

47. John Mahon notes for example that "since there had been several harvest failures, the province was suffering from real hunger." See J. K. Mahon, *The War of 1812* (Gainsville, FL: University of Florida Press, 1972), 258.

48. *A Narrative of the Life, Travels and Adventures of Captain Israel Adams: Who lived at Liverpool, Onondaga County, New York.* http://www.villageofliverpool.org/public_ftp/IsraelAdamsStory.pdf.

49. "Original Letters," *The Balance and State Journal (1809–1811)* 1, no. 14 (April 2, 1811): 106.

50. In later life, Kerr moved back to his birthplace in Albany, New York. Newspaper accounts indicate that Kerr was highly regarded in both countries. John Ross Robertson, *The History of Freemasonry in Canada from its foundation in 1749*, 2 vols. (Toronto, Ont.: George Morang and Company, Limited, 1900) 1:481; Cuyler Reynolds, *Albany Chronicles: A History of the City Arranged Chronologically* (Albany, NY: J. B. Lyon Company, Printers, 1906), 442. Campbell discusses Clinton's wish to see Dr. Kerr while in Niagara, Upper Canada in *Life and Writings of De Witt Clinton*, 124 and 127.

51. Charles G. Roland, "KERR, ROBERT," in *Dictionary of Canadian Biography*, vol. 6, University of Toronto/Université Laval, 2003– , accessed March 31, 2017, http://www.biographi.ca/en/bio/kerr_robert_6E.html.; Isabel T. Kelsay, "TEKARIHOGEN (1794–1832)," in *Dictionary of Canadian Biography*, vol. 6, University of Toronto/Université Laval, 2003– , accessed March 31, 2017. http://www.biographi.ca/en/bio/ tekarihogen_1794_1832_6E.html.

52. This point is substantiated in Peter Grande, "The Political Career of Peter Buell Porter, 1797–1829" (PhD diss., University of Notre Dame, 1971), 26.

53. Grande, "The Political Career of Peter Buell Porter," 28.

54. Address before the House of Representatives, February 1810, in *Annals of Congress*, 11th Congress, 2nd session, 1392. Porter's estimates and plans for the Niagara Canal were taken chiefly from Albert Gallatin's report. Peter B. Porter Papers, February 17, 1810, Roll 11, Misc.

55. In 1811 for example, T. Dickson, T. Clark, and T. Cummings of Niagara, Upper Canada received scythes, books, and other items from Porter, Barton and Co. "Permit for unloading the cargo of the *ONTARIO*," June 17, 1811, Augustus Porter Papers, Box 7, Folder 3, 200.52. The Buffalo History Museum. See also Emily Cain, "Provisioning Lake Ontario merchant schooners, 1809–1812: Lord Nelson(Scourge), Diana(Hamilton), Ontario and Niagara," *Freshwater* 3, no. 1 (Summer 1988): 21–25. Porter also had regular business contacts with merchants

in Upper Canada. See n.61. Emily Cain writes that the Porter vessel the *Niagara* was suspected of smuggling goods from Upper Canada immediately before the war. Cain, "Customs Collection—and Dutiable Goods: Lake Ontario Ports, 1801–1812," *Freshwater* 2, no. 2 (Autumn, 1987): 24. In later years, Porter, Barton and Co. reached out to fellow tourist developers on the Canadian side of the River in an effort to promote roads and other internal improvements at Niagara Falls. See, for example, *The Buffalo Patriot*, reprinted in "Article 3—No title," *The New York Mirror: A Weekly Gazette of Literature and Fine Arts* vol. 3, no. 8 (September 17, 1825): 59. For more on Porter and Tourism see chap. 6.

56. See, for example, Merritt's letter to Porter, Barton and Co., dated December 1816. Merritt Family Papers [microform], MG 24 E1, Brock University Archives. As will be discussed further in chap. 4, Merritt also spoke to Porter during the canal age about the possibility of expanding the ship lock at Porter's home town of Black Rock on the Niagara River for Canadian use. See chap. 4, n.123. Merritt also wrote a letter to Porter about a proposed regulation allowing American wheat and flour into Canada duty free. Letter to Augustus Porter from Peter Porter, April 9, 1831, Augustus Porter Papers, Folder 2, 600.046. The Buffalo History Museum.

57. J. C. A. Stagg, "Between Black Rock and a Hard Place: Peter B. Porter's Plan for an American Invasion of Canada in 1812," *Journal of the Early Republic* 19, no. 3 (Autumn, 1999): 388. For more on this subject see chap. 3, n.62.

58. Shaw, *Erie Water West*, 40.

59. O. Turner, *History of the Pioneer Settlement*, 613.

60. *Annals of Congress*, 12th Congress, 1st session, February 18, 1812, 1059. Historian Peter Grande wrote "The people of the Niagara Frontier were badly divided . . . on foreign affairs and a loud chorus of protest arose rejecting the wild Anglo-phobic declaration of war made in Washington." Grande, "The Political Career of Peter Buell Porter," 30.

61. Letter reprinted in the *Buffalo Gazette*, April 21, 1812.

62. Just over a quarter of a century later, a poignant statement in the *Army and Navy Chronicle* announced that General Peter B. Porter had the honor of reviewing a British regiment at Niagara Falls, Upper Canada, "in the very vicinity of the spot where he stood" during the late war. "Review of a Regiment of British Troops by Gen. P. B. Porter," *Army and Navy Chronicle* (1835–1842) 9, no. 5 (August 1, 1839): 71. For more on Porter and his relationship to the "War Hawks" see Stagg, "Between Black Rock and a Hard Place," 386. Alan Taylor also writes that while Porter vacillated on the question of war with Canada, he did pursue one constant goal: "to protect his mansion, wharves, and stores, which lay within cannon shot of the British." Alan Taylor, *The Civil War of 1812*, 132.

63. Elizabeth McKinsey discusses the origin of the honeymoon at Niagara Falls in McKinsey, *Niagara Falls*, 178–82. For the reference to Morris and his wife see Jack Kelly, *Heaven's Ditch: God, Gold, and Murder on the Erie Canal* (New York: St. Martin's Press, 2016), 16. See also Koeppel, *Bond of Union*, 80.

64. Campbell, *Life and Writings of De Witt Clinton*, 135.

65. Porter's estimates for the canal were chiefly taken from the Report. Peter B. Porter Papers, February 17, 1810, Roll 11, Misc. The Buffalo History Museum. Albert Gallatin, *Report of the Secretary of the Treasury; on the subject of Roads and Canals* (Washington DC: Printed By Order of the Senate, 1810), 47.

66. Campbell, *Life and Writings of De Witt Clinton*, 133–34.

67. Campbell, *Life and Writings of De Witt Clinton*, 129–30.

68. See for example Koeppel, *Bond of Union*, 84.

69. Campbell, *Life and Writings of De Witt Clinton*, 140. Despite the impending crisis, the canal commissioners, before returning to New York City, chose to convene in Chippewa, Upper Canada, where they discussed their findings.

70. Shaw, *Erie Water West*, 42; McGreevy, *Stairway to Empire*, 22.

71. This is especially true of popular accounts. Peter Bernstein for example writes that the commissioners "quickly reached agreement" that the canal would run overland between Lake Erie and the Hudson, thereby avoiding the Lake Ontario route. Bernstein, *Wedding of the Waters*, 154–56. The argument that the interior route was inevitable even as early as 1810 is also hinted at in Kelly, *Heaven's Ditch*, 8–11.

72. Clinton is believed to have used the pseudonym *Tacitus* to write on the Erie Canal. See *The Canal Policy of the State of New York; Delineated in a letter to Robert Troup, Esquire*. By Tacitus (Albany, NY, 1821), 24.

73. Mc Nall, *Agricultural History*, 98–100.

74. This point is also discussed in William Charles Lahey, "The Influence of David Parish on the Development of Trade and Settlement in Northern New York, 1808–1822" (PhD Diss., Syracuse University, 1958), 42.

75. Shaw, *Erie Water West*, 42.

76. Stuart, *United States Expansionism*, 70 and 110; Errington, *The Lion, the Eagle and Upper Canada*, 37; Sheppard, *Plunder, Profit, and Paroles*, 155; McCalla, *Planting the Province*, 31.

77. *Canal Laws*, 1:63–64.

78. As quoted in *Niles Weekly Register*, April 30, 1814.

79. *Canal Laws*, 1:75.

80. *Canal Laws*, 1:51. For more on the connections between Porter and the Michigan officials see Ronald Shaw, "Michigan Influences upon the Formative Years of the Erie Canal," *Michigan History* 37, issue 1 (March 1953), 3–5 and 13–14.

81. *Canal Laws*, 1: 104–05.

82. *Buffalo Gazette*, November 15, 1814.

83. *Buffalo Gazette*, March 4, 1815.

84. *The Canal Policy of the State of New York; Delineated in a Letter to Robert Troup, Esquire*, 40.

85. Nathan Miller, *The Enterprises of a Free People: Prospects of Economic Development in New York State During the Canal Period, 1792–1838* (Ithaca, NY:

Cornell University Press, 1962), 40; Evan Cornog, *The Birth of Empire: De Witt Clinton and the American Experience, 1769*–*1828* (Oxford, UK: Oxford University Press, 1998), 116; Shaw, *Erie Water West*, 57.

86. *The Canal Policy of the State of New York; Delineated in a Letter to Robert Troup, Esquire*, 29.

87. Published in the *Buffalo Gazette*, August 13, 1816.

88. Ronald Shaw, *Erie Water West*, 143.

89. *Kingston Gazette*, October 13, 1818.

90. *Buffalo Gazette*, February 27, 1816.

91. *The Canal Policy of the State of New York; Delineated in a Letter to Robert Troup, Esquire*, 34.

92. Jane Errington notes that Upper Canadians were anxious to quickly reestablish peace after 1815 and restore contacts, both formal and informal, with the neighboring United States that were vital to the province's well-being. Errington, *The Lion, the Eagle and Upper Canada*, 120. Reginald Stuart, though interested in the later theme of manifest destiny, acknowledges that after 1815 the British provinces were important outlets for the borderland American economy in this region. Stuart, *United States Expansionism*, 83.

93. *Buffalo Gazette*, June 13, 1815.

94. *Buffalo Gazette*, June 20 and August 15, 1815.

95. *Buffalo Gazette*, June 6, 1815.

96. *Canal Laws*, 1:294. Printed also in the *Buffalo Gazette*, January 7, 1817.

97. Speech reprinted in the *Buffalo Gazette*, February 13, 1816; Robert Troup, Esq., *A Letter to the Hon. Brockholst Livingston*, 17.

98. *Buffalo–Black-Rock Harbor Papers*, 41–42; Cornog, *The Birth of Empire*, 116.

99. *Canal Laws*, 1:118; *Buffalo Gazette*, March 26, 1816.

100. Nathan Miller, *The Enterprises of a Free People*, 61–62.

101. *Canal Laws*, 1:118; *Buffalo Gazette*, March 26, 1816.

102. *Kingston Gazette*, February 1, 1817.

103. Errington, *The Lion, the Eagle and Upper Canada*, 148.

104. *Kingston Gazette*, September 7, 1816.

105. The question of the complementary New York and Upper Canadian Canal system will be expanded upon in chapters 4 and 5.

106. Shaw, *Erie Water West*, 78.

107. *Kingston Gazette*, March 1, 1817. See also *Kingston Gazette*, February 1, March 15, and April 12, 1817. *Kingston Chronicle*, January 29 and February 26, 1819 and January 28, 1820.

108. Gideon Granger, Esq. "Speech Delivered before a Convention of the People of Ontario County, New York . . . on the Subject of a Canal from Lake Erie to the Hudson River." (Canandaigua, NY: 1817).

109. Shaw, *Erie Water West*, 74.

110. Charles Haines, *Considerations on the Great Western Canal* (New York: Spooner & Worthington Printers, 1818), 16.

111. Shaw, *Erie Water West*, 114–115.

112. *Kingston Gazette*, October 6, 1818.

113. *Kingston Chronicle*, February 26, 1819.

114. *Kingston Chronicle*, March 5, 1819.

115. *Kingston Chronicle*, May 2, 1823.

116. The reverse was also true with hundreds of American laborers, both skilled and unskilled, seeking employment opportunities on Canada's Welland Canal. Expanded on in chap. 4.

117. *Kingston Gazette*, September 29, 1818.

118. In addition to their presence on the Oswego Canal, as will be discussed in a later chapter, American canal builders on the Illinois and Michigan line employed dozens of Canadians. See Kathleen R. Arnold, ed., *Contemporary Immigration in America: A State by State Encyclopedia* (Santa Barbara, CA: Greenwood, an imprint of ABC-CLIO, LLC, 2015), 258; Jim Redd, *The Illinois and Michigan Canal* (Carbondale and Edwardsville, IL: Southern Illinois University Press, 1993), 55; Way, *Common Labour*, 100, n.59.

119. *Kingston Chronicle*, November 12, 1819.

120. See for example *Upper Canada Herald*, November 9, 1819; *Kingston Chronicle*, January 28, 1820, August 24, 1821, and November 23, 1821.

121. See chap. 3, n.100.

122. *Kingston Chronicle*, January 28, 1820.

123. *Upper Canada Herald*, October 21, 1823.

124. *Niagara Gleaner*, July 8, 1824.

125. Shaw, *Erie Water West*, 181–94; Koeppel, *Bond of Union*, 365–86; Errington, *The Lion, the Eagle and Upper Canada*, 119.

126. Koeppel, *Bond of Union*, 228;

127. Shaw, *Erie Water* West, 185–86.

128. John Rutherford, *Facts and Observations in Relation to the Origin and Completion of the Erie Canal* (New York: H. B. Holmes, 1825), 18. Courtesy of the The Buffalo History Museum. See also George Condon, *Stars in the Water: The Story of the Erie Canal* (New York: Doubleday, 1974), 174.

129. The Upper Canadian *Niagara Gleaner* was fascinated by the fact that the "sound of the cannonading on the completion of the Grand Canal was heard at Albany 3 minutes before 11." For a reference to the celebration see the *Upper Canada Herald*, November 15, 1825, and *Kingston Chronicle*, November 11, 1825.

130. Jane Errington notes that the celebration was commented upon by the *Upper Canada Herald* in November 1825. Errington, *The Lion, the Eagle, and Upper Canada*, 119. For the most complete account of this story see Cadwallader D. Colden, *Memoir Prepared at the Request of a Committee of the Common Council of the City of New York, and Presented to the Mayor of the City, at the Celebration of*

the Completion of the New York Canals (New York, 1825), 322. The details of the Canal Celebration can also be read in its entirety in the "Canal Celebration," *The New York Mirror: A Weekly Gazette of Literature and the Fine Arts (1823–1842)* 3, no. 16 (November 12, 1825): 126.

131. *Buffalo Emporium*, November 12, 1825.

132. *Niagara Gleaner*, December 18, 1824.

133. "Miscellaneous Items," *The New England Farmer, and Horticultural Register (1822–1890)* 4, no. 20 (December 9, 1825): 159.

134. *Farmers Journal and Welland Canal Intelligencer*, October 11, 1826.

135. "Summary," *The Friendly Visitor, being a Collection of Select and Original Pieces, Instructive and Entertaining, Suitable to be Read in all Families (1825–1825)* 1, no. 46 (November 16, 1825): 368.

136. Bukowczyk et al., *Permeable Border*, 34.

137. In particular, Tucker notes that these advantages came after the American Drawback Laws that gave Canadians a free choice of markets for their goods. Gilbert Tucker, *The Canadian Commercial Revolution* (Toronto, Ont.: McClelland and Stewart Limited, 1964, reprinted 1970), 155. See also Bukowczyk et al., *Permeable Border*, 41.

138. *Black Rock Gazette*, June 15, 1826.

139. *Black Rock Gazette*, November 24, 1827.

140. American packet-boat companies regularly advertised in Upper Canada's *Farmer's Journal and Welland Canal Intelligencer*. See for example, September 6, June 7, and September 13, 1826.

141. *Niagara Gleaner*, July 8, 1824.

142. *Montreal Courant*, reprinted in The *Farmer's Journal and Welland Canal Intelligencer*, July 19, 1826.

143. *Farmer's Journal and Welland Canal Intelligencer*, January 17, 1827.

144. "Abridged Report of the Board of Directors of the Welland Canal Company, for 1831." *Canadian Emigrant* (Sandwich, Ont.), February 4, 1832.

145. Ronald Shaw observes that the Erie Canal brought "an altogether new and stimulating experience" in travel to "the first canal generation in New York." However, it is worth noting that Upper Canadians were equally enthralled by the new travel opportunities in the neighboring Republic. Shaw, *Erie Water West*, 197. Indeed, no one more valued the new and stimulating experience of boarding an Erie or Oswego canal packet than William Hamilton Merritt who regularly traveled to the United States on canal business.

146. *Niagara Gleaner*, September 24, 1824. For more on how Clinton was admired in Canada see Errington, *The Lion, the Eagle and Upper Canada*, 124–25.

147. *Black Rock Gazette*, November 23, 1826.

148. *Upper Canada Herald*, December 5, 1826.

149. *Buffalo Patriot*, September 29, 1829.

150. After 1828, the Oswego Canal was also advertised by the Canada Company as an inexpensive means of transport to the upper province. During official

business in North America, Lord Durham spoke of the considerable number of immigrants in the 1830s "who entered the province by way of New York and the Erie Canal." Sir C. P. Lucas, *Lord Durham's Report on the Affairs of British North America*, 3 vols. (Oxford, UK: Oxford at the Clarendon Press, 1912), 2:217. See also *A Statement of the Satisfactory Results which have Attended Emigration to Upper Canada, From the Establishment of the Canada Company until the Present* (London: Smith, Elder and Co., 1841), 57; *Canadian Emigrant* (Sandwich, Ont.), December 1, 1834; John Galt, *The Autobiography of John Galt in Two Volumes* (Philadelphia, PA: Key and Biddle, 23 Minor Street, 1834) 1:205–206; Clarence Karr, *The Canada Land Company: The Early Years* (Toronto, Ont.: Ontario Historical Society Research Publication, 1974), no. 3: 10 and 13 and 34–35 and 85–87.

151. Charles Z. Lincoln, *Messages of the Governors*, 11 vols. (New York: J. B. Lyon Press, 1909) 2:1011.

152. *Onondaga Register*, November 2, 1825. See also Cadwallader Colden, *Memoir Prepared at the Request of the Committee of the Common Council of the City of New York at the Celebration of the Completion of the New York Canals*, 303.

153. Hosack, *Memoir of De Witt Clinton*, 73.

154. Colden, *Memoir Prepared at the Request of a Committee of the Common Council of the City of New York, and Presented to the Mayor of the City, at the Celebration of the Completion of the New York Canals*, 285; George W. Holley, *The Falls of Niagara and other Famous Cataracts* (New York: A. C. Armstrong and Son, 1883), 99–100.

155. For more on Nathan Roberts and the Lockport Locks see McGreevy, *Stairway To Empire*, 33 and 43 and 51 and 61–62 and 100 and 128; Eric Lawson, *Nathan Roberts: Erie Canal Engineer* (Utica, NY: North Country Books, Inc., 1997); Koeppel, *Bond of Union*, 332–34.

156. The report of Nathan Roberts, Civil Engineer, January 1826, can be read in *Report on the location and expense of a ship canal around Niagara Falls: also, from the Illinois River to Lake Michigan: with a report of a select committee to the Assembly, April 14, 1834, relating to the connection from Oswego to the Hudson* (New York: Published at the Office of the Railroad Journal, 1834). Courtesy of The Buffalo History Museum.

157. *Niles Weekly Register*, October 22, 1826.

Chapter 4

1. "Traveller, A," *Western Recorder (1824–1833)*, October 17, 1826.

2. The First Welland Canal was formerly opened in November of 1829, but a shorter and more direct route to Lake Erie was completed in 1833.

3. Creighton, *The Empire of the St. Lawrence*, 196–97; Aitken, *The Welland Canal Company*, 22; Styran and Taylor, *This Great National Object*, 42; Jackson, *The Mighty Niagara*, 131; Shaw, *Erie Water West*, 413; Koeppel, *Bond of Union*, 118.

4. The Oswego Canal is analyzed more fully in chap. 5.

5. See, for example, Victor Konrad, "Borderlines and Borderlands in the Geography of Canada-United States Relations," in J. Randall and Herman Konrad, *North America Without Borders: Integrating Canada, the United States, and Mexico* (Calgary, Alberta: University of Calgary Press, 1992), 191; Stuart, *United States Expansionism*, 109; Landon, *Western Ontario and the American Frontier*, vii.

6. Donald C. Masters, "Evolution of a Frontiersman," *The Manitoba Arts Review* 2, no. 1 (Spring 1940): 54–55; and J. J. Talman, "William Hamilton Merritt," *The Loyalist Gazette* 13, no. 1 (Spring 1975): 2–3; Alan Taylor, *The Civil War of 1812*, 206.

7. Merritt's journal of events while a prisoner in the United States can be read in William Wood, editor, *Select British Documents of the Canadian War of 1812*, 3 vols. (Toronto, Ont.: Greenwood Publishing Group, 1928) 3, Part 2: 623–24. A familiar presence along the Niagara border, Dr. Cyrenius Chapin of Buffalo, New York, practiced medicine on both sides of the border before the war. Reginald Stuart notes that Dr. Chapin settled first in Canada and worked both sides of the Niagara River so that by 1810 "a borderland emerged in the Niagara region," in Stuart, *United States Expansionism*, 44. See also n.9 below.

8. Aitken, *The Welland Canal Company*, 25.

9. As with Dr. Chapin, Prendergast represented a "striking number of men who practiced this profession in the province" indicating that a cross-border community had emerged early in the Niagara–Great Lakes region. Historian Fred Landon writes that before 1840 the American contribution to medicine in Upper Canada "was noteworthy." See Landon, *Western Ontario and the American Frontier*, 53.

10. Merritt wrote several letters to his future American wife while a prisoner in the United States; these letters can be read online. See Ontario Ministry of Government Services titled "Prisoners of War." www.archives.gov.on.ca/en/explore/online/1812/prisoners.

11. For more on Dr. Prendergast see *The Centennial History of Chautauqua County* (Jamestown, NY: The Chautauqua History Co., 1904), 115; Helen G. McMahon, *Chautauqua County: A History* (Buffalo, NY: Henry Stewart Inc., 1958), 39; Obed Edson, *Biographical and Portrait Cyclopedia of Chautauqua County* (Philadelphia, PA: Published by John M. Gresham and Co., 1891), 72; Aitken, *The Welland Canal Company*, 25–26 and 54; Styran and Taylor, *This Great National Object*, 3 and 20.

12. Merritt subscribed jointly to the New York newspaper with George Keefer whose son, Thomas Coltrin Keefer apprenticed as an engineer on the Erie Canal before being appointed provisional engineer on the Welland Canal. Aitken, *The Welland Canal Company*, 43. See chap. 7. n.7 for more on T. C. Keefer.

13. Aitken writes that Merritt's first vision for water improvements had to do with the constant shortage of water on his property in connection with his mills. Aitken, *The Welland Canal Company*, 28–29.

14. Aitken, *The Welland Canal Company*, 29.

Notes to Chapter 4   209

15. William H. Merritt to J. Prendergast, St. Catharines, September 3, 1818, Merritt Family Papers [microform], MG 24 E1, Brock University Archives; and J. P. Merritt, *Biography*, 53. Later scholarship indicates that from the start, finances, or lack of Canadian capital, emerged as a major problem during the building of the Welland Canal. See Aitkin, *The Welland Canal Company*, 47; Creighton, *Empire of the St. Lawrence*, 270; Styran and Taylor, *This Great National Object*, 24–25.

16. See for example, "Traveller, A," *Western Recorder (1824–1833)*, October 17, 1826. See also the *St. Catharines Journal*, June 2, 1842.

17. For more on Merritt's admiration of New York State see Donald C. Masters, "Evolution of a Frontiersman," *The Manitoba Arts Review* 2, no. 1 (Spring 1940): 54–59.

18. For more on the Welland Survey see *Transactions of the Canadian Society of Engineers* 11, Part 1, January–June (Montreal, Quebec: Printed for the Society, 1888): 41. Information on Tibbett(s)'s survey can also be found in the *Farmer's Journal and Welland Canal Intelligencer*, June 6, 1827. See also J. N. Jackson, *The Welland Canal and Their Communities, Engineering, Industrial, and Urban Transformation* (Toronto, Ont.: University of Toronto Press, 1997), 35.

19. *Appendix to the Journal of the House of Assembly of Upper Canada*, Session 1836–7, vol. 2: *Titled No. 90 Third Report From the Select Committee* (Toronto, 1836), 213. Hereafter cited *Third Report*.

20. *Third Report*, 217.

21. *Niagara Gleaner*, March 12, 1825.

22. George Keefer, President of the Welland Canal Company *Circular*, St. Catharines, June 28, 1824, in the First Welland Canal file, St. Catharines Museum and Welland Canal Center, St. Catharines, Ont. Also in *Third Report*, 223.

23. *Colonial Advocate*, September 27, 1824.

24. Merritt met up with Clinton in Albany during one of his many visits to the capital while on canal business. See J. P. Merritt, *Biography*, 80.

25. *Tribute to the Memory of De Witt Clinton, Late Governor of the State of New York* (Albany, NY, 1828), 128. Quote lamenting Clinton's death in Upper Canada in *Farmer's Journal and Welland Canal Intelligencer*, February 27, 1828. This quote can also be read in Errington, *The Lion, the Eagle and Upper Canada*, 125.

26. W. H. Merritt to Catharine Merritt, October 18, 1824. Merritt Family Papers [microform], MG 24 E1, Brock University Archives.

27. Reprinted in "Political," *Boston Recorder (1817–1824)* 9, no. 44 (October 30, 1824): 175.

28. *Third Report*, 251.

29. *The New York Daily Advertiser* reprinted in "Political." *Western Recorder (1824–1833)*, November 7, 1826: 179.

30. Shaw, *Erie Water West*, 413, n. 41.

31. Miller, *The Enterprises of a Free People*, 107. Miller clarifies that England was the largest investor in the Canal fund.

32. For more on the British and American investors in the Erie Canal see Miller, *The Enterprises of a Free People*, 99–111; Koeppel, *Bond of Union*, 210–11.

33. Reference to the British shareholders can be found in the "Summary," *Christian Register (1821–1835)* 7, no. 25 (June 21, 1828): 99. See also Aitken, *The Welland Canal Company*, 88.

34. For the names of the stockholders see *Third Report*, 286–87. In 1826, a woman by the name of Elizabeth Ball was also awarded property damages by the Welland Canal Company. See *Farmer's Journal and Welland Canal Intelligencer*, September 13, 1826. For a more general history on the role of women in transportation see Beatrice Craig, *Women and Business Since 1500: Invisible Presences in Europe and North America* (New York: Palgrave Macmillan, 2015).

35. These costs were estimated by Merritt for the period 1824–1836. See Merritt, *Brief Review of the Origin, Progress, Present State, and Future Prospects of the Welland Canal* (St.Catharines: H. Leavenworth, 1852), 21. Hereafter, Merritt, *Brief Review*.

36. There is no biography on Yates but references to the American financier can be found in a variety of sources. See for example "Will of the Late John. B. Yates," *Niles Weekly Register*, August 6, 1836; Hugh Aitken, "Yates and McIntyre: Lottery Managers" *Journal of Economic History* 13, no. 1 (Winter 1953): 36–57; Aitken's *The Welland Canal Company*, probably has the most extensive information on Yates.

37. Landon, *Western Ontario and the American Frontier*, 51.

38. John B. Yates file, Historian's Office, Town of Sullivan Office Building, Chittenango, NY; J. P. Merritt, *Biography*, 162–63; "Will of the Late John B. Yates," reprinted in *Black Rock Advocate*, August 4, 1836.

39. Aitken, *The Welland Canal Company*, 79; McCalla, *Planting the Province*, 169.

40. Shaw, *Erie Water West*, 413. Merritt's biographer writes that in addition to Yates who invested in the project, Alfred Hovey, one of many Erie Canal contractors on the Welland, took $10,000 of stock. J. P. Merritt, *Biography*, 63.

41. *Papers Respecting Claim of Shareholders in the Late Welland Canal Company, for Arrears in Interest, Under the Act 7th Victoria, Chapter 34* (Quebec: Printed by Rollo Campbell, 1853), 3–4.

42. *Third Report*, 472. See also J. P. Merritt, *Biography*, 85.

43. *Third Report*, 354.

44. Yates remained a devout supporter of the Welland Canal until his death in 1836, despite the financial hardships that it brought to him personally. See Aitken, *The Welland Canal Company*, 116–18; Jack Williams, *Merritt: A Canadian Before His Time* (St. Catharines, Ont.: Stonehouse, 1985), 40; McCalla, *Planting the Province*, 169–70; Creighton, *The Empire of the St. Lawrence*, 300.

45. J. P. Merritt, *Biography*, 63.

46. Merritt's speech is reprinted in the *Upper Canada Gazette*, January 13, 1825.

47. *Upper Canada Gazette*, January 13, 1825.

48. Reprinted in the *Niagara Gleaner*, December 4, 1824.

49. New York's interest in the Welland Canal was indicated by the fact that Boulton's lengthy speech was reprinted in its entirety in the *Black Rock Gazette*, November 9, 1826. The speech can also be read in *Report from the Select Committee on Emigration from the United Kingdom* (Ordered by the House of Commons to be Printed, May 26, 1826), 13–22.

50. Historian Gilbert Tucker writes that "it is a well-known principle of transportation economics that up to a certain point efficiency increases with the size of the unit." The same author recognized the advantage of the St. Lawrence River and Great Lakes shipping paths that had fewer tolls than the Erie Canal, in Tucker, *The Canadian Commercial Revolution*, 38.

51. *Detroit Gazette*, December 5, 1826. Courtesy of the Maritime History of the Great Lakes digital library.

52. *Buffalo Emporium*, June 25, 1825.

53. "Art. XXII—Account of the Welland Canal, Upper Canada," *American Journal of Science and Arts (1820–1879)* 14, no. 1 (July, 1828): 159.

54. Upper Canada was said to be disadvantaged particularly following the opening of the Oswego Canal that drew trade from Lake Ontario into the Erie Canal system at Syracuse. For a discussion of how the New York canal system challenged Canadian interests see Thomas McIlwraith, "Transport in the Borderlands," in Lecker, *Borderlands: Essays in Canadian-American Relations*, 62 and Thomas McIlwraith, "Freight Capacity and Utilization of the Erie and Great Lakes Canals Before 1850," *Journal of Economic History* 36, no. 4 (December 1976): 856–76; Shaw, *Erie Water West*, 414; Tucker, *The Canadian Commercial Revolution*, 43–44; Creighton, *The Empire of the St. Lawrence*, 251.

55. McCalla, *Planting the Province*, 121–22.

56. Bukowczyk et al., *Permeable Border*, 41.

57. J.P. Merritt, *Biography*, 51.

58. William Hamilton Merritt, *Brief Review*, 34–40.

59. For example, Alexander Yates McDonnell was a nephew of John B. Yates (the largest private investor in the Welland Canal), serving as vice president of the Welland Company and director. William E. Lansing was also from New York but moved to Upper Canada to act as company director for two years before returning to the United States. For more on the American presence on the Canadian canal see Styran and Taylor, *This Great National Object*, 97–100.

60. *Annual Report of the Welland Canal Company for 1832* (St. Catharines: H. Leavenworth, 1833), 10.

61. *Third Report*, 253.

62. The Report of the President and Directors of the Welland Canal Company to the Stockholders (York: U.C.: Printed By Robert Stanton, 1826), 11.

63. For more on Benjamin Wright and James Geddes see Richard Weingardt, *Engineering Legends: Great American Civil Engineers* (Reston, VA: American Society of Engineers, 2005), 4–9; William H. Shank, *Towpaths to Tugboats: A History of American Engineering* (York, PA: The American Canal and Transportation Center, 1982), 19–22; David S. Landes et al., eds, *The Invention of Enterprise: Entrepreneurship from Ancient Mesopotamia to Modern Times* (Princeton, NJ: Princeton University Press, 2012), 341–42; Benjamin Hall Wright, *Origin of the Erie Canal* (Rome, NY: Sandford and Carr, Book and Job Printers, 1870); Whitford, *History of the Canal System of the State of New York*, 2: chap. 3; Koeppel, *Bond of Union*, 6–8 and 8–9.

64. *Black Rock Beacon*, June 29, 1826.

65. Shaw, *Erie Water West*, 413.

66. As stated, an impressive cadre of European and American born engineers assisted on the Welland Canal. Francis Hall, a Scottish-born engineer who trained under the renowned British Engineer Thomas Telford, emigrated to Canada in 1823, likely to find employment on the Canadian project. Samuel Clowes, a British engineer who gained experience constructing canals in Europe and the United States, was also called upon to consult on the Canadian project, but increasingly the company turned to the neighboring United States for help in building their canal.

67. *Third Report*, 226. The *Buffalo Journal* also announced that contracts were being let on the canal and that these positions were being advertised in the *Utica Sentinel*, the *Albany Gazette*, the *Rochester Telegraph*, and the *Lockport Observatory* among other papers. *Buffalo Journal*, April 26, 1825.

68. *Rochester Telegraph* reprinted in the *Niagara Gleaner*, November 6, 1824.

69. A local Buffalo newspaper noted that Merritt also visited the Erie Canal at Utica, New York, where he met with contractors "with a view to prosecute the work immediately." *Buffalo Emporium and General*, October 23, 1824. For the reference to Merritt's visit to Rochester see J. P. Merritt, *Biography*, 57.

70. *New York Spectator* reprinted in the *Buffalo Emporium*, October 23, 1824.

71. *Third Report*, 227 and 235; J. P. Merritt *Biography*, 72.

72. John N. Jackson, "Four Welland Canals," Paper read before the Canadian Canal Society and The Canal Society of New York State, St. Catharines, Ontario, September 25–27, 1992; and Robert Finch, *Story of the New York State Canals: Historical and Commercial* (State of New York, State Engineer and Surveyor, 1925), 11.

73. J. P. Merritt, *Biography*, 72.

74. "Art. XXII—Account of the Welland Canal, Upper Canada," *American Journal of Science and Arts (1820–1879)* 14, no. 1 (July, 1828): 159.

75. As quoted in the *Colonial Advocate*, May 12, 1825.

76. Phelps's decision to contract on the Welland Canal is detailed in the *Third Report*, 50–51.

77. McGreevy, *Stairway to Empire*, 128.

78. *In Memoriam: Orson Phelps, Esq.* (Portland, OR: Geo. H. Himes, Printer, 1870), 10–11. See reference to Oliver Phelps in Perry Smith, *History of the City of Buffalo and Erie County*, 2 vols. (Syracuse, NY: D. Mason and Co., Publishers, 1884), 2:64.

79. Styran and Taylor note that by 1825 at least seven of the firms on the First Welland Canal had worked on the Erie Canal. Styran and Taylor, *This Great National Object*, 99 and 101–02.

80. Way, *Common Labour*, 12–13.

81. There is little research concerning labor on the original or First Welland Canal (1824–1833). Labor history on the Welland Canal largely relates to the period of the 1840s after the First Welland Canal was completed.

82. *Third Report*, 84 and 91.

83. See, for example, an ad placed by Phelps seeking workers from south of the border with their animals. *Farmer's Journal and Welland Canal Intelligencer*, July 25, 1827. The Welland Canal Company also advertised work opportunities in several American newspapers. See, for example, *Buffalo Emporium*, October 23, 1824; *Black Rock Gazette*, November 1, 1825; *Buffalo Journal*, June 19, 1827; and the *Black Rock Gazette*, September 29, 1827.

84. *New York Commercial Advertiser* reprinted in *Niles Weekly Register*, October 22, 1824.

85. *Rochester Telegraph* as cited in "Miscellaneous," *The Columbian Star (1822–1829)* 4, no. 26 (June 25, 1825): 102.

86. *Niagara Gleaner*, September 24, 1825.

87. *Niagara Gleaner*, November 5, 1825.

88. *Farmers Journal and Welland Canal Intelligencer*, May 2, 1827.

89. Hansen, *The Mingling*, 1:103.

90. Way, *Common Labour*, 77.

91. *Buffalo Journal* on June 19, 1827.

92. *Farmer's Journal and Welland Canal Intelligencer*, July 11, 1827.

93. McGreevy, *Stairway to Empire*, 65.

94. *Third Report*, 381.

95. *Third Report*, 80–81.

96. J. P. Merritt, *Biography*, 117.

97. For Charlotte Phelps's account see *The Women's Standard*, May 23, 1895. There is little written on labor relations on the First Welland Canal. For relations between labor and contractor on all of the North American canals see Way, *Common Labour*, 37 and 68 and 112. On Canada's Second Welland Canal during the 1840s see Styran and Taylor, *This Great National Object*, 283. On the Erie Canal see McGreevy, *Stairway to Empire*, 73 and Sheriff, *The Artificial River*, 42.

98. The advertisement is reprinted in Roberta Styran and Robert R. Taylor, *Mr. Merritt's Ditch* (Erin, Ont.: A Boston Mills Press Book, 1992), 100.

99. *Farmer's Journal and Welland Canal Intelligencer*, July 25, 1827. For more on woman and the family on all of the North American canals see Way, *Common Labour*, 84–85 and 112 and 171–72.

100. McGreevy, *Stairway to Empire*, 71.

101. In 1827, the Welland Company promised rewards to "Every overseer" who could ensure "peace and good order" on the line. Another statement that Roman Catholic laborers had been subject to "ill usage from persons of other persuasions" offers a bleak insight into the life of canal workers. *Third Report*, 381.

102. *Farmer's Journal and Welland Canal Intelligencer*, June 13, 1827. Peter Way observes that such cases of violence and death on the canal lines rarely received more than a passing notice in the newspapers. "Violence and murder, it was believed, were natural to canal work sites." Way, *Common Labour*, 178–79.

103. *Third Report*, 380. Witnessing the labor discipline on the Welland line during an 1827 visit, Basil Hall wrote "all the parties seem to know what they are about; there is . . . no idleness; no remission." From the "Colonial," *The Albion, A Journal of News, Politics, and Literature* (1822–1876), 6, no. 11 (August 25, 1827), 85.

104. *Farmer's Journal and Welland Canal Intelligencer*, May 30, 1827.

105. *The Annual Report of the Board of Directors of the Welland Canal, for 1827* (St. Catharines: H. Leavenworth, 1828), 3; *Third Report*, 364.

106. "Art. XXII—Account of the Welland Canal, Upper Canada," *American Journal of Science and Arts (1820–1879)* 14, no. 1 (July, 1828): 159.

107. *Farmer's Journal and Welland Canal Intelligencer*, July 25, 1827.

108. *Third Report*, 380.

109. From the *St. Catharines Journal*, reprinted in the *Kingston Chronicle*, Nov. 29, 1828.

110. James Geddes to Peter B. Porter, January 21, 1829, Peter B. Porter Papers, Roll 11, Misc. The Buffalo History Museum. The Geddes and Barrett reports detailing the alternate route may be read in the *Third Report*, 383–84 and 399.

111. William H. Shank, *Towpaths to Tugboats: A History of American Canal Engineering*, 38; and Styran and Taylor, *Mr. Merritt's Ditch*, 34.

112. Douglas McCalla observes that governments on both sides of the lakes could impose policies that at times sought to tax or inhibit some trades: "But people, goods, and information" still moved both ways despite these polices. McCalla, *Planting the Province*, 7.

113. *Third Report*, 246.

114. Contractors went wherever materials were most cost efficient and readily available. During the building of the Buffalo Harbor at the western end of the Erie Canal for example, contractor Samuel Wilkeson of Buffalo purchased stone from across the Niagara River near Fort Erie to save money. As told by Captain James Sloan, "A Pioneer Trader with an Account of his Share in the Building of the Buffalo Harbor," Buffalo and Erie County Historical Society *Publications*

(Buffalo, New York, 1902) 5: 236. Reference to Split Rock quarry from Way, *Common Labour*, 77.

115. Several of McGuire's countrymen also found work on the Oswego and other New York and Canadian canals. Theresa Bannan, *Pioneer Irish of Onondaga, About 1776–1847*. (New York: G. P. Putnam's Sons, 1911), 121 and 176 and 180–81.

116. *Black Rock Gazette*, September 29, 1827. A year earlier, the *Black Rock Beacon* wrote, "It must be gratifying to our citizens, to have the fact stated, that all the principal engineers, who assisted in constructing the Erie canal, are eagerly sought after, . . . in the neighboring province and several of the states. Mr. engineer Thomas is now employed on the Welland Canal." *Black Rock Beacon*, June 29, 1826.

117. David I. Spangel, *De Witt Clinton and Amos Eaton: Geology and Power in Early New York* (Baltimore, MD: Johns Hopkins University Press, 2014), 132.

118. *Niles Weekly Register*, October 31, 1829.

119. *Niles Weekly Register*, December 26, 1829. As earlier indicated, smuggling was widespread in the Niagara–Great Lakes region and the New York and Upper Canadian canal system facilitated this illegal trade. In 1829, the editor of the *Niagara Gleaner* wrote "the temptation to smuggle on both sides is so great; so much so, that an honest trader cannot live; he must smuggle in self-defense." From the *Niagara Gleaner* reprinted in the *Detroit Gazette*, August 13, 1829. Courtesy of the Maritime History of the Great Lakes digital library. Smuggling from Canada to the Erie Canal was reported during the same time in the *Kingston Gazette and Religious Advocate*, November 13, 1829.

120. *New York Spectator*, November 27, 1829.

121. *Buffalo Republican*. Reprinted in the *Detroit Gazette* (Detroit, MI), December 17, 1829, 2. Courtesy of the Maritime History of the Great Lakes digital library.

122. Richard Garrity, *Canal Boatman: My Life on Upstate Waterways* (Syracuse, NY: Syracuse University Press, 1977), 147.

123. Regarding the regulation on American grains see Letter from Peter B. Porter to Augustus Porter, April 9, 1831 in Augustus Porter Papers, Box 11, Folder 2, 600.046. Merritt's letter to Peter Porter regarding the improvement of the Black Rock lock is available in print in Robert J. A. Irwin, Jr., "William Hamilton Merritt and the First Welland Canal," *Niagara Frontier* 2, no. 4 (Winter 1964): 115–16.

124. *Buffalo Republican*. Reprinted in the *Detroit Gazette* (Detroit, MI), December 17, 1829, 2. Courtesy of the Maritime History of the Great Lakes digital library.

125. *Annual Report of the Welland Canal Company for 1829*, 3.

126. J. P. Merritt, *Biography*, 123.

127. As recounted in the *Oswego Palladium*, July 14, 1830.

128. *Kingston Chronicle*, January 2, 1830.

129. Aitken, *The Welland Canal Company*, 65.

130. The new line shortened the canal from 33 to 28 miles. *Montreal Gazette*, June 20, 1833. Courtesy of the Maritime History of the Great Lakes digital library.

131. "Article 4—no Title," *The Albion, A Journal of News, Politics and Literature (1822–1876)* 8, no. 28 (December 19, 1829): 223. A year earlier, one visitor to Buffalo wrote that "the Welland, or some other projected canal, threatens to deprive the village of its advantages" but "I have no doubt, that the unrivalled advantage" of the village, "will secure to it a triple ratio of increase, when the latter canal shall be completed." "A Tour," *The Western Monthly Review* 2, no. 5 (October 1828): 249.

132. "Article 1—no Title," *The Banner of the Constitution. Devoted to General Politics, Political Economy, State Papers, Foreign and Domestic News, &c. (1829–1832)* 1, no. 4 (December 26, 1829): 32.

133. Edwin Williams, *The New York Annual Register For the Year . . . 1830* (New York: J. Leavitt Publisher, 1830–1834), 128–29.

134. Ronald Shaw, "Michigan Influences upon the Formative Years of the Erie Canal," 15.

135. Annual Report of the Welland Canal Company for 1829, 5–6.

136. *Buffalo Republican*, August 15, 1829.

137. Bukowczyk et al., *Permeable Border*, 198, n.48. One Canadian scholar notes that the Welland Canal did more to "swell the Oswego Canal than to supply Montreal, thereby placing both sides of the Niagara Frontier within the commercial hinterland of New York City." Jackson, *The Mighty Niagara*, 138. This issue will be discussed further in chap. 5.

138. New York State Legislative Documents, 53rd session, vol. 4, 1830. The Canal Board was responding to the diversion of American produce through Canada when tolls got too high on the Erie. See *Niles Weekly Register*, June 12, 1830; *New York Spectator*, June 25, 1830.

139. *Third Report*, 472.

140. From the *Albany Evening Journal*, reprinted in the *Onondaga Register and Syracuse Gazette*, April 7, 1830.

141. "Article 8—no Title," *Christian Secretary (1822–1889)* 12, no. 33 (August 31, 1833): 131.

142. *Niagara Gleaner*, July 6, 1833.

143. Glazebrook observes that the competition of the Welland Canal lowered tolls on the Erie Canal that was a benefit to both sides. Glazebrook, *A History of Transportation* 2vols. (Toronto, Ont.: McClelland and Stewart Limited, 1964), 1:88. As noted earlier, Canadian historian Gilbert Tucker writes that this was especially evident after the passage of the American Drawback Laws: "An increasing benefit of having competing routes was a gradual reduction of tolls" allowing farmers and merchants to use whatever route was cheapest at a given time. Tucker, *The Canadian Commercial Revolution*, 155.

144. "Article 1—no Title," *The Banner of the Constitution. Devoted to General Politics, Political Economy, State Papers, Foreign and Domestic News, &c. (1829–1832)* 1, no. 4 (December 26, 1829): 32.

145. *New York Spectator*, May 11, 1830.

146. *New York Spectator*, September 3, 1830.

147. *Buffalo Republican*, December 12, 1829.

148. Reprinted in the *Kingston Chronicle*, October 9, 1830.

149. Jackson, *The Mighty Niagara*, 139; "Literary and Philosophical Intelligence, etc," *The Christian Advocate (1823–1834)* 9 (July 1, 1831): 371.

150. From a Report of the Directors of the Welland Canal Company in 1831 in *Canadian Emigrant* (Sandwich, Ont.), February 9, 1832. Courtesy of the Maritime History of the Great Lakes digital library.

151. Henry Cook Todd, *Notes upon Canada and the United States, from 1832 to 1840: Much in a small space, or a great deal in a little book* (Toronto, Ont.: Printed by Rogers and Thompson, 1840), 94. Online Text. Retrieved from the Library of Congress, https://www.loc.gov/item/01027841/ (accessed April 02, 2017).

152. "Remniscences of a Recent Journey from Cincinnati to Boston," T. F. *The Knickerbocker, or New York Monthly Magazine* (1833–1862), October 24, 1833, 242.

153. *Niagara Gleaner*, November 2, 1833.

154. Shaw, *Erie Water West*, 266. The rise of Rochester as a vital international port on Lake Ontario both before and after the Erie Canal can be read in Joan Sullivan, "Entrepreneurship in the Genesee Country," *Rochester History* 45 no(s). 3 and 4 (July and October, 1983), 1–24. See also chap. 4, n. 156.

155. The Oswego Canal also became an indispensable link in the inland line of communications between New York and Upper Canada.

156. COMMUNICATIONS: Statistics of Rochester, *The Genesee Farmer and Gardener's Journal (1831–1839)* 5, no. 48 (November 28, 1835): 380. Shaw also states that wheat arriving from the west by way of the Welland Canal also stopped at Rochester "on the way to Montreal." Shaw, *Erie Water West*, 266.

157. "Report of the Canal Commissioners," *Hazard's Register of Pennsylvania, Devoted to the Preservation of Facts and Documents, and Every Kind of Useful Information Respecting the State of Pennsylvania (1831–1835)* 13, no. 2 (January 11, 1834): 28.

158. *Report of the Committee of the Commons House of Assembly of Upper Canada, Relative to the Welland Canal, with the Letter of John B. Yates, Esq.* (St. Catharines, Ont., 1834), 19.

159. "Welland Canal," *The Albion, A Journal of News, Politics and Literature (1822–1876)* 3, no. 7 (February 14, 1835): 55.

160. *Buffalo Whig and Journal*, June 11, 1835.

161. *Buffalo Whig and Journal*, June 11, 1835.

162, "Article 1—no Title," *The New Yorker (1836–1841)* 10, no. 8 (November 7, 1840): 126. As indicated earlier, scholars debate the long-term advantage of the Welland Canal when compared to the Erie. See chap. 1, n.17.

163. Abridged Report of the Welland Canal Company for 1831, reprinted in the *Canadian Emigrant*, February 9, 1832. Courtesy of the Maritime History of the Great Lakes digital library.

164. *Niles Weekly Register*, Nov. 27, 1830.

165. "Article 8—no Title," *Christian Secretary* (1822–1889) 12, no. 33 (August 31, 1833): 13. The same article noted that a reduction of tolls on the Erie Canal would ameliorate some of the losses to the Canadian canal.

166. From the *Cleveland Herald*, reprinted in the *British American Journal*, July 29, 1834.

167. From the *Cleveland Herald* reprinted in the *St. Catharines Journal*, May 24, 1838.

168. *Third Report*, 420.

169. "Methodist Episcopal Church," *The New Yorker (1836–1841)* 9, no. 13 (June 13, 1840): 207.

170. "Michigan," *Atkinson's Saturday Evening Post (1831–1839)* XIII, no. 609 (March 30, 1833): 1.

171. Stuart, *United States Expansionism*, 119.

172. *Kingston Chronicle and Gazette*, May 20, 1835.

173. *Oswego County Whig*, May 1, 1839. See also *St. Catharines Journal*, May 16, 1839. See also Richard Palmer, Getting "stoned" on the Welland Canal, *Inland Seas* 47, no. 3 (1991) at http://images.maritimehistoryofthegreatlakes.ca/results?sort=madePublic+desc&itype=Articles&fz=0.

174. "Art. XXII—Account of the Welland Canal, Upper Canada," *American Journal of Science and Arts (1820–1879)* 14, no. 1 (July, 1828): 159.

Chapter 5

1. Reference in "Advertisement 1—no Title," *New York Farmer (1828–1837)* 10, no. 11 (July 30, 1837): 256.

2. *The Western Guidebook and Emigrant's History: Containing General Descriptions of Different Routes Through the State of New York, Ohio, Indiana, Illinois, and Michigan* (Buffalo, NY: Oliver G. Steele Publisher, 1834).

3. According to this view, the Oswego Canal was a strategic measure on the part of the United States to steer trade coming off the Welland Canal and Lake Ontario back toward New York and the Hudson River. At the same time it was argued that the Oswego Canal, in conjunction with the Welland, drew freight toward American ports that might have otherwise traveled through the Canadian Canal system. Koeppel, Bond of Union, 84–85; McIlwraith, "Transport in the Borderlands," in Robert Lecker, ed., *Borderlands: Essays in Canadian-American Relations*, 61–62; Larkin, *New York State Canals*, 49–50 For a view of the Oswego Canal's role within the context of the Canada–United States borderland see Bukowczyk et al., *Permeable Border*, 41.

4. Charles Snyder, *Oswego: from Buckskins to Bustles* (Port Washington, NY: Ira J. Friedman, 1968), 83. Because there are no book-length studies on the Oswego Canal, scholarly analysis of this vital waterway is limited.

5. In Hosack, *Memoir of De Witt Clinton*, 269.

6. See chap. 3 n. 152. Noble Whitford writes "The Erie Canal, as finally constructed, passed through Syracuse south of Onondaga lake, but both during and after its construction the agitation continued for the connection of the waters of the Erie Canal and Lake Ontario at Oswego." Whitford, *History of the Canal system of the State of New York*, 1: 446.

7. *Canal Laws*, 1:450.

8. *Kingston Chronicle*, November 2, 1819.

9. "Article 1—no Title," *The Plough Boy (1819–1820)* 1, no. 32 (January 8, 1820): 254.

10. The reference to the Porter brothers promoting the Oswego Canal(as an adjunct to a Niagara Canal) can be read in a letter from Porter, Barton and Co. partner Benjamin Barton to Peter B. Porter, Lewiston, March 6, 1820, Roll 11, Misc. Peter B. Porter Papers, The Buffalo History Museum. The reference to the Salina side-cut is in the Journal of the Senate of the State of New York, 1824, 47th session, 312. See also Snyder, *Oswego*, 86.

11. *Kingston Chronicle*, December 22, 1822, as quoted in Errington, *The Lion, the Eagle and Upper Canada*, 123.

12. Announced in the *Wayne Sentinel* (Palmyra), October 29, 1823.

13. Letter of Gerrit Smith to the public in *Onondaga Register and Syracuse Gazette*, November 3, 1830.

14. "An Act to incorporate the Oswego Canal Company," *Oswego Palladium*, June 27, 1823.

15. Whitford, *History of the Canal System of the State of New York*, 1:448.

16. For the reference to Bronson as an international business man see Emily Cain, "Customs Collection—and Dutiable Goods," 22. For Bronson's commercial connections to Porter see Augustus Porter Papers, Box 4, Folder 3, 200.052. The Buffalo History Museum. See also *Oswego Palladium*, February 3, 1877. Latter courtesy of the Maritime History of the Great Lakes digital library. This commercial connection was early announced in a local newspaper advertisement: "Messrs. Townsend, Bronson & Co. at Oswego and Porter, Barton & Co. at Lewiston, have made such arrangements as will facilitate the transportation of Goods, Salt, etc." *Buffalo Gazette*, October 15, 1816. For Alvin Bronson's biography and death notice see the *New York Times*, June 23, 1878 and April 3, 1881.

17. *Oswego Palladium*, August 14, 1824. Daniel Larkin writes that "The Oswego Canal was second only to the Erie in number of boat clearances (an indication of canal usage) during the 25 years after it opened." Larkin, *New York State Canals*, 50. However, as indicated later in this chapter, the Oswego Canal's success was also tied to Canada's Welland Canal.

18. *Kingston Chronicle*, March 11, 1825.

19. Errington, *The Lion, the Eagle and Upper Canada*, 121.

20. *Oswego Palladium*, August 7, 1824. There is very little information on workers on the private Oswego Canal for this period. Theresa Bannan briefly details

the experiences of Irish workers on the Oswego Canal following its assumption by the State. Bannan, *Pioneer Irish of Onondaga*, 121 and 126 and 193. For the experience of immigrant labor on the early, mainly private canals, see Peter Way, *Common Labour*, 18–46. For more on this subject see chap. 5, n.21.

21. Life for the Scottish workers however was apparently not easy, it being observed in one account that in 1823 the Oswego Company "kept us months and months without pay." Edward Shortt, *Perth Remembered* (Perth, Ont.: Printed by Mortimer Ltd., 1967), 30–31; Michael Vance, *Imperial Immigrants: Scottish Settlers in the Upper Ottawa Valley, 1815–1840* (Toronto, Ont.: Dundurn Press, 2012), 143; Robert W. Passfield, *Military Paternalism: Labor and the Rideau Canal Project* (Bloomington, IN: Author House, LLC). 95, n.123.

22. Edwards was also an abolitionist and Gerrit Smith's agent at Oswego. His home in Oswego was a well-known way station on the Underground Railroad. Crisfield Johnson, *History of Oswego County, New York* (New York: L. H. Everts and Company, 1877), 187.

23. "An Act to Incorporate the Oswego Canal Company, April 23, 1823," can be read in its entirety in the *Oswego Palladium*, June 27, 1823. Carol Sheriff's skillfully crafted account of the Erie Canal reminds us that internal improvements often came at a high cost to the land and the people. Sheriff, *The Artificial River*, 79.

24. The report continued that there was a large class of people engaged in the business and that "more than 1000 barrels of eels were caught annually, with 500 barrels of other fish, at the Oswego Falls." *Canal Laws*, 1:501. See also *Documentary Sketch of the New York State Canals by S. H. Sweet, Deputy State Engineer and Surveyor. Accompanying State Engineer and Surveyor's Report for* 1862 (Albany, NY: VanBenthuysen Printer, 1863), 111. An account was also told of the thriving salmon fishery which found its way "to a foreign market" but as a result of the damming of the river by the State, the fish "have not visited the waters of the Oswego." Joseph V. H. Clark, *Onondaga; or Reminiscences of Earlier and Later Times; Being a Series of Historical Sketches Relative to Onondaga . . . and Oswego*, 2vols. (Syracuse, NY: Stoddard and Babcock, 1849) 2:388.

25. For the reference to property damage on the Welland see *Niagara Gleaner*, December 14, 1826. See also chap. 3, n.51 reference to John Brant who represented his people's interests on the Welland Canal in Isabel T. Kelsay, "TEKARIHOGEN (1794–1832)," in *Dictionary of Canadian Biography*, vol. 6, University of Toronto/ Université Laval, 2003.

26. "Petition for connecting Lake Ontario with the Erie Canal," quoted in the *Oswego Palladium*, December 13, 1823.

27. "Petition for connecting Lake Ontario with the Erie Canal," quoted in the *Oswego Palladium*, December 13, 1823.

28. *A Gazetteer of the State of New York: Embracing An Ample Survey and Description of Its Counties, Towns, Villages, Canals . . .* , by Horatio Gates Spafford, LL.D (Albany, NY: 1824), 387.

29. Despite State assumption of the canal, the private Oswego Canal Company continued to possess rights to the surplus waters that fed its hydraulic canal.

30. Journal of the Assembly of the State of New York, 1825, 48th session, 687.

31. Cole Harris and John Warkentin, *Canada Before Confederation: A Study of Historical Geography* (Kingston: McGill-Queen's University Press, 2000), 151.

32. This story is recounted in Debbie Stack, *The Best from American Canals*, no. 3 (York, PA: The American Canal and Transportation Center, 1989), 12. See also Snyder, *Oswego*, 87.

33. *Canal Laws*, 2:233.

34. Journal of the Senate of the State of New York, 1826, 49th session, 432.

35. "Annual Report of the New York Canal Commissions," Journal of the Assembly of the State of New York, 1827, 50th session, 503.

36. Journal of the Senate of the State of New York, 1826, 49th session, 353.

37. Way, *Common Labour*, 69.

38. *Salina Sentinel*, April 11, 1827. During the same time the *New York Enquirer* advertised that "stone cutters, etc, on the Rideau Canal" were needed. Reprinted in the *Kingston Chronicle*, April 6, 1827. Like much writing on the Canadian canal system, conflict and rivalry has dominated the narrative of the Rideau Canal that linked the Ottawa River and Lake Ontario, and was originally designed, in the War of 1812 aftermath, to bypass the American border on the St. Lawrence. However, when viewed from the perspective of the northern borderland, the story of the Rideau is more nuanced. At a brief glance, the porosity of the international border is evident during the Rideau's building. The Rideau's original director, Colonel John By, for example, advertised in the United States for contractors and laborers to come to the Canadian line. For the American presence on the Rideau Canal see William N. T. Wylie, "Poverty, Distress, and Disease: Labour, and the Construction of the Rideau Canal, 1826–1832," *Labour/Le Travail* 11 (Spring 1983): 4. See also Robert Legget, Rideau Waterway (Toronto, Ont.: University of Toronto Press, 1955), 218; Merridee L Bujaki, "Parallel Challenges Building the New York and Rideau Canals," *International Journal for the History of Engineering and Technology* 84, no. 1 (January 2014): 30–51. For the Rideau as a military response to the United States after the war see Jackson, *The Mighty Niagara*, 116; Styran and Taylor, *This Great National Object*, 35.

39. Quoted in the *Black Rock Beacon*, June 9, 1827.

40. Though Peter Way does not analyze the Oswego Canal, he observes that wages on many of the North American lines varied based on the season, the difficulty of the job, the level of skill required, or the danger of the work involved. A worker's wages were also determined by such variables as age, gender, race, and ethnicity. Way, *Common Labour*, 111–12. Lionel Wyld speaks briefly to wage variables in *Low Bridge! Folklore and the Erie Canal* (Syracuse, NY: Syracuse University

Press, 1962), 19 and 21. I know of no studies that speak directly to wages on the Oswego Canal.

41. *Journal of the Senate of the State of New York*, 1826, 49th session, 253–55.

42. For a biographical sketch of the engineers see Whitford, *History of the Canal System of the State of New York*, 2: chap. 3, "Biographies of Engineers." See also Shaw, *Canals for a Nation*, 161–64; Styran and Taylor, *Mr. Merritt's Ditch*, 38. Peter Way is one of the few historians who offers a comparative framework when analyzing engineers and contractors in Canada and the United States. Way, *Common Labour*, 12–13.

43. *Salina Sentinel*, January 31, 1827.

44. *Journal of the Assembly of the State of New York*, 1829, 52nd session, 188.

45. Ibid, 188.

46. *Salina Sentinel*, August 22, 1827.

47. Phelps' letter was written to the editor of the *Upper Canada Herald* on August 25, 1827 but published on September 4, 1827. Styran and Taylor observe that by October much sickness and death pervaded the Welland Canal. Styran and Taylor, *This Great National Object*, 279.

48. "Article 1—no Title," *Niles' Weekly Register (1814–1837)* 35, no. 895 (Nov 08, 1828): 161.

49. *Buffalo Journal*, June 19, 1827.

50. *Journal of the Assembly of the State of New York*, 52nd session, 1829, 188.

51. New York *Daily Advertiser*, reprinted in *Niles Weekly Register*, November 8, 1828.

52. Whitford, *History of the Canal System of the State of New York*, 1:452.

53. *Niles Weekly Register*, May 30, 1829.

54. *St. Catharines Journal*, August 20, 1830. Previous quote from *Oswego County Fifty Years Ago. Address Delivered By Hon. R. H. Tyler before the Old Settlers' Association, at Mexico, Thursday August 21, 1879* (Fulton, NY: Morrill Bros. Printing, 1880), 5.

55. Snyder, *Oswego*, 91; Larkin, *New York State Canals*, 50; John C. Churchill, editor, *Landmarks of Oswego County New York* (Syracuse, NY: D. Mason and Company Publishers, 1895), 158.

56. *Onondaga Register and Syracuse Gazette*, May 20, 1829.

57. *Onondaga Register and Syracuse Gazette*, May 5, 1830.

58. *Onondaga Register and Syracuse Gazette*, October 20, 1830.

59. American ads can be read in the *Kingston Gazette and Religious Advocate*, October 9 and November 6, 1829; *Upper Canada Herald*, September 23 and October 14, 1829; *Patriot's Farmer's Monitor* (Kingston), October 12, 1829.

60. *Oswego Gazette* reprinted in the *Farmer's Journal and Welland Canal Intelligencer*, July 11, 1827.

61. J. P. Merritt, *Biography*, 127.

62. *Oswego Gazette* reprinted in the *Wayne Sentinel* (Palmyra, NY), September 23, 1829. According to John Churchill, "one of the first consequences" of the opening of the Oswego Canal, and the anticipated completion of the Welland Canal, "was the incorporation of the village" of Oswego in 1828. Churchill, ed., *Landmarks of Oswego County, New York*, 158.

63. *Free Oswego Press*, March 24, 1830.

64. From the *Kingston Chronicle* reprinted in the *Oswego Palladium*, June 2, 1830. Chrisfield Johnson notes that a surge of interest in Oswego's milling industry followed on the heels of the Welland Canal's opening. Johnson, *History of Oswego County*, 149.

65. John C. Churchill, ed., *Landmarks of Oswego County New York*, 332.

66. Merritt, *Biography*, 317.

67. *Records of the Village of Oswego, 1828–1848* (Oswego, NY: Oswego, Daily and Weekly Times Print, 1874), 60. Records can be read at the Library Annex, Cornell University.

68. *Oswego Palladium*, August 8, 1830.

69. The *Oswego Palladium* noted Smith's presence at the occasion stating he "was then, fortunately, on a visit to this place, and whose large investments in capital, and munificent expenditure in the improvement of our village, creates a deep interest in everything affecting the fortunes of this village." *Oswego Palladium*, August 11, 1830.

70. *Free Oswego Press*, August 11, 1830.

71. *Upper Canada Herald*, August 25, 1830. See also Canada's *Colonial Advocate*, September 2, 1830, for a favorable review of the occasion.

72. *Colonial Advocate*, September 2, 1830.

73. *Oswego Palladium*, October 20, 1830.

74. Reprinted in the *Oswego Palladium*, October 20, 1830.

75. Joshua V. H. Clark, *Onondaga; or Reminiscences of Earlier and Later Times*, 2:390.

76. *Oswego Palladium*, May 23, 1832.

77. *Oswego Palladium*, June 12, 1833. Reginald Stuart observes that in 1841 businessman James Tallmadge of New York "noted the growth of the borderland trade and emphasized its value to the United States. He and others pointed to Oswego as an example of a lake port flourishing because of provincial markets." Stuart, *United States Expansionism*, 118.

78. Snyder, *Oswego*, 94.

79. *St. Catharines Journal*, June 1, 1837.

80. Snyder, *Oswego*, 104–05.

81. *Canadian Emigrant and Western District Advertiser*, December 1, 1831. Courtesy of the Maritime History of the Great Lakes digital library.

82. *Kingston Chronicle*, July 23, 1831.

83. From the *Oswego Palladium*, reprinted in the *Kingston Chronicle*, June 11, 1831.

84. *British Whig*, November 14, 1834.

85. Hamilton also owned the *Great Britain*. For more on Hamilton see Peter Baskerville, "HAMILTON, JOHN, (1802–82)," in *Dictionary of Canadian Biography*, vol. 11, University of Toronto/Université Laval, 2003– , http://www.biographi.ca/en/bio/hamilton_john_1802_82_11E.html accessed April 2, 2017; Frank Mackey, *Steamboat Connections: Montreal to Upper Canada, 1816–1843* (Montreal and Kingston: Mc Gill-Queens University Press, 2000), 146–47. See also *Kingston Chronicle and Gazette*, March 24, 1841.

86. Delays were often caused by congestion and breaches in the canal. In 1826, several memorials from western New York were read before the Senate complaining "of a want of water in the Erie Canal" during various times of the year. Peter B. Porter Papers, entitled "In Senate, March 21, 1826," Roll 11, Misc. The Buffalo History Museum. Complaints of breaches at the western end of the canal were also reported in the *Buffalo Patriot*, reprinted in the *St. Lawrence Gazette* (Ogdensburg, NY), June 13, 1826. Courtesy of the Maritime History of the Great Lakes digital library. In 1827 the *Buffalo Patriot* reported that the canal would likely be closed for ten to twelve days due to damage at the western end. *Buffalo Patriot*, reprinted in "Domestic." *Western Recorder* (1824–1833), April 24, 1827. In St. Catharines, Upper Canada, it was reported during the same time that "Much delay and vexation have resulted from the multiple breaches in the canal, [Erie] since it was opened this spring. Up to last Saturday, at noon, only an occasional boat has reached us from Albany." In *Farmer's Journal and Welland Canal Intelligencer*, May 19, 1830. See also Koeppel, *Bond of Union*, 259 and 264 and 282–83.

87. *Oswego Free Press*, reprinted in the *Oswego Palladium*, March 31, 1830.

88. *Oswego Free Press*, reprinted in the *Farmer's Journal and Welland Canal Intelligencer*, October 30, 1830.

89. From the *Oswego Palladium*, reprinted in the *British Whig*, September 1, 1835.

90. See for example *British Whig*, February 11 and 25, 1834. *Kingston Chronicle and Gazette*, May 20, 1835.

91. From the Kingston *Oswego Palladium* reprinted in the *Kingston Chronicle*, November 14, 1835.

92. *Kingston Chronicle and Gazette*, October 28, 1835.

93. *Oswego Palladium*, July 29, 1835.

94. Creighton, *The Empire of the St. Lawrence*, 251. One historian noted that Canadian opponents of the Welland Canal often claimed that it had been built mainly for the benefit of Oswego, since it paid to bring the larger cargoes of the

lake craft (filled with wheat and flour) through it and transship it to Oswego for the Erie Canal. A. R. M. Lower, *The North American Assault on the Canadian Forest: A History of the Lumber Trade between Canada and the United States* (New York: Greenwood Press, 1968), 116, n.39. Discussing the growing trade and resulting prosperity between Oswego and Upper Canada, Reginald Stuart writes "Oswego flour mills ground Canadian wheat for shipment to Kingston and down the St. Lawrence." Stuart, *United States Expansionism*, 118. Jackson, *The Mighty Niagara*, 137–38. For more on the question of the Oswego canal as a contender for the Canadian trade see chap. 1, n.17 and chap. 5, n.100.

95. *Oswego Palladium*, June 17, 1835. Newspapers on both sides of the border brought attention to the growing trade. Typical was an ad placed in the Oswego Palladium stating: "Thus by the Oswego and Welland Canal, Ohio wheat has been received and manufactured into flour and in all probability will be delivered in New York—and also goods consigned from New York for the same route, will have been received in Cleveland before the ice is out of the harbor of Buffalo." *Oswego Palladium*, reprinted in the *Kingston Chronicle and Gazette*, May 20, 1835.

96. Creighton, *The Empire of the St. Lawrence*, 252.

97. *Upper Canada Herald*, May 10, 1836. Lower, *The North American Assault*, 116; Stuart, *United States Expansionism*, 122.

98. *Kingston Chronicle and Gazette*, October 28, 1835.

99. *Upper Canada Herald*, April 14, 1835.

100. Recognizing the complexity of the issue, Bukowczyk writes "At a more intermediate spatial level, the Welland Canal carried grain to Montreal, but at the same time (thanks to the Oswego branch of the Erie) transformed both Hamilton and Toronto into termini of the Erie system, a funnel through which goods might enter but, conversely, also a funnel through which staples might leave Upper Canada." Bukowczyk et al., *Permeable Border*, 41.

101. Creighton, *The Empire of the St. Lawrence*, 252.

102. *Oswego Palladium*, June 17, 1835.

103. Not only was Smith one of the largest realty holders in Oswego, having purchased much of the land bordering the Oswego Canal, but he was also a proponent of open and free trade with Canada. The reference to Smith's realty investments comes from the *Oswego Palladium*, August 11, 1830. On Smith's, and Bronson's, views on reciprocity, the Welland Canal, and trade ties to Canada see *Speeches of Gerrit Smith in Congress* (New York: Mason Brothers, 1855), 316; Gerrit Smith letter to F. Whittlesey, Peterboro, February 1, 1845, in Gerrit Smith Broadside and Pamphlet Collection, Syracuse University, Special Collections Online. http://scrconline.syr.edu/xtf/view?docId=metsmods/smith_g_pb.70795.metsmods.xml;brand=scrc. See also Octavius B. Frothingham, *Gerrit Smith, A Biography* (New York: G. P. Putnam Sons, 1879), 32; Alvin Bronson, *An Essay on the Commerce and Transportation of the Vallies* [sic] *of the Great Lakes and Rivers, of the North-west* (Oswego, NY: Advertiser and Times Steam Press, 1868); Johnson, *History of Oswego County, New*

York, 75 and 152; The Lumber Era (1840–1928)—Port of Oswego, NY, OSWEGO HISTORIAN. http://oswegohistorian.org/2010/09/the-lumber-era-1840-1928-port-of-oswego-ny/. For Merritt and Reciprocity in Canada see Tucker, *The Canadian Commercial Revolution*, 101–02. For the argument that reciprocity would lead to annexation by the United States, Gerrit Smith argued "free trade will supersede" any desire for annexation. Smith like other Americans believed that free trade would extinguish annexationist desires in the United States. *Speeches of Gerrit Smith in Congress*, 317. For more on annexation see Bukowczyk et al., *Permeable Border*, 49.

104. As discussed earlier, I know of no studies that consider canals and reform as a cross-border phenomenon in this region. For a discussion of American religious and benevolent influence on Upper Canada see Fred Landon, *Western Ontario and the American Frontier*, 134–38; Meinig, *The Shaping of America*, 2: 51.

105. "Traveller, A," *Western Recorder (1824–1833)*, November 29, 1825.

106. *Letters on the Natural History and Internal Resources of the State of New York by Hibernicus* [De Witt Clinton]. (New York, 1822), 25.

107. Carol Sheriff writes that those at the forefront of the canal age would struggle "to find ways to make compatible the various meanings of improvement and progress." Sheriff, *The Artificial River*, 25, 139. For a discussion of the historiography of reform during the transportation and market revolutions in the United States see Patrick McGreevy, *Stairway to Empire*, 182–85. On the Canada side, Styran and Taylor briefly consider the reform agenda on all four Welland Canals. Styran and Taylor, *This Great National Object*, 266–68.

108. *The American Tract Magazine for the Year 1828*. (New York: D. Fenshaw Printer, 1828) 3, no. 10 (October 1828): 122.

109. The reference to "equal privileges and equal protection" indicated the early presence of religious conflict and abuse on the Canadian canal. *Third Report*, 380–81. Despite the efforts of the company, in later years religious strife and violent outbreaks plagued the Welland and other North American lines. Some of the worst of this religious strife and violence occurred in neighboring Lockport on the Erie Canal. McGreevy, *Stairway to Empire*, 70–71.

110. J. A. Tuer, *An Historical Narrative of Some Important Events in the Life of First Church, St. Catharines, 1831–1931* (Toronto, Ont.: United Church Publishing House, 1931), 13–15; "Article 4—no Title," *The Friend; a Religious and Literary Journal (1827–1906)* 2, no. 50 (September 26, 1829): 396.

111. "Extracts From The Diary Of A Mission Among The Indians At New-Fairfield, (Upper Canada), May, 1830, To April, 1831," *The United Brethren's Missionary Intelligencer, and Religious Miscellany; Containing the most Recent Accounts Relating to the United Brethren's Missions among the Heathen; with Other Interesting Communications from the Records of that Church (1822–1848)* 4, no. 9 (1833): 415. See also chap. 2, n.8.

112. "Cause of Seamen," *The Religious Intelligencer . . . Containing the Principal Transactions of the various Bible and Missionary Societies, with Particular Accounts of Revivals of Religion (1816–1837)* 15, no. 27 (December 4, 1830): 422.

113. "American Seamen's Friend Society, *Episcopal Recorder (1831–1851)* 10, no. 8 (May 26, 1832): 31.

114. "American Seamen's Friend Society," *Christian Secretary (1822–1889)* 12, no. 17 (May 11, 1833): 65. See also an earlier article in the *Kingston Chronicle* that gives attention to the "Lake Ontario Seamen's Friend Society," July 31, 1830.

115. "American Tract Society," *New York Evangelist (1830–1902)* 7, no. 3 (January 16, 1836): 11.

116. "Art. V—Seamen's Friend Society," *Quarterly Christian Spectator (1829–1838)* 2, no. 2 (June 1, 1831): 253.

117. "American Seamen's Friend Society," *Christian Secretary (1822–1889)* 13, no. 17 (May 10, 1834): 66. While I know of no book-length studies on the subject of reform on either of the Welland or Oswego canals, Carol Sheriff points out in her study of the Erie Canal that the Bethel reformers' "efforts to improve the moral conditions of boatmen met with uneven success. Some workers showed considerable interest, while others were outright hostile. . . . While some years brought noticeable improvement in Canal morality, subsequent years brought back sliding." Sheriff, *The Artificial River*, 155.

118. *Third Report*, 26.

119. By 1828, three years after the Erie Canal's opening, Buffalo, New York, was known for its large number of drinking establishments that catered to thousands of sailors, canalers, boatmen, and travelers who passed through the town. David A. Gerber, *The Making of an American Pluralism: Buffalo, New York 1825–60* (Urbana and Chicago, IL: University of Illinois Press, 1989), 26. Phelps also witnessed much alcohol abuse while contracting on the Erie Canal at Lockport. For the use of alcohol at the Lockport worksite see McGreevy, *Stairway to Empire*, 76–79.

120. *The Women's Standard*, May 23, 1895. See chap. 5, n.122.

121. The *Intelligencer* made the latter comment about the "reformation" in relation to the native population near the Welland Canal. "Progress of Temperance." *The Religious Intelligencer . . . Containing the Principal Transactions of the various Bible and Missionary Societies, with Particular Accounts of Revivals of Religion (1816–1837)* 14, no. 17 (September 19, 1829): 270.

122. Canadian historians Styran and Taylor believe that employers like Phelps "made attempts as they could—some enlightened, some misguided—to alleviate living and working conditions." Styran and Taylor, *This Great National Object*, 264–65. Other scholars view such measures more skeptically. For the impact of alcohol on labor/contractor relations see Peter Way, *Common Labour*, 183, and McGreevy, *Stairway to Empire*, 79.

123. This section of Hall's journey of the Welland Canal is also reprinted in the *Black Rock Gazette*, September 29, 1827.

124. A much-sought-after professional, Barrett distinguished himself on several North American projects. Working on both sides of the international border, Barrett supervised the building of New York's Chenango Canal, and the Erie Canal's enlargement after 1837. More information on Alfred Barrett can be found

in Larry McNally, "BARRETT, ALFRED," in *Dictionary of Canadian Biography*, vol. 7, University of Toronto/Université Laval, 2003– , http://www.biographi.ca/en/bio/barrett_alfred_7E.html, accessed April 2, 2017. See also Whitford, *History of the Canal System of the State of New York*, 2:1148.

125. Peter Way writes that by 1845 there were 124 unlicensed grog shops (or five per mile) on the Welland Canal. Way, *Common Labour*, 182.

126. *British Whig* (Kingston), May 22, 1835.

127. *Kingston Chronicle*, July 31, 1830.

128. *Kingston Chronicle*, July 30, 1830.

129. "American Seamen's Friend Society," *Christian Secretary (1822–1889)* 13, no. 17 (May 10, 1834): 66.

130. See for example, the *Kingston Chronicle*, May 5 and 12 and 19 and June 2, 1832.

131. Frothingham, *Gerrit Smith*, 29 and 114. See also chap. 5, n.103.

132. "Miscellaneous," *New York Observer and Chronicle (1833–1912)* 11, no. 49 (December 7, 1833): 196.

133. Johnson, *History of Oswego County*, 187; Frothingham, *Gerrit Smith*, 113–14. Snyder, *Oswego*, 100–01; Lisa Brenda, "Oswego Celebrates its Role in the Underground Railroad," *Oswego County Today* (September 14, 2010). See also Snodgrass, *The Underground Rail Road: an encyclopedia of people, places, and* operations, 2 vols. (New York: M. E. Sharpe, 2008) 1:120 and 227; 2:495.

134. Mary Ellen Snodgrass, *The Underground Railroad* 1:99 and 120 and 363–64; Jackson, *The Mighty Niagara*, 155–57.

135. Landon, *Western Ontario and the American Frontier*, 134–43.

136. From the *Kingston Patriot* reprinted in the *Oswego Palladium and Chronicle*, September 1, 1830. Colden first mentions the value of the floating museums to education in *Memoir Prepared at the Request of a Committee of the Common Council of the City of New York, and Presented to the Mayor of the City, at the Celebration of the Completion of the New York Canals*, 88.

137. See Samuel Rezneck, "A Travelling School of Science on the Erie Canal in 1826," *New York History* 40, no. 3 (1959): 255–69; "Domestic." *Western Recorder (1824–1833)*, May 30, 1826.

138. From the *Kingston* Patriot (Upper Canada)t reprinted in the *Oswego Palladium*, September 1, 1830. For more on this subject see David Spanagel, *DeWitt Clinton and Amos Eaton*, 123–27.

139. In 1828, the *American Journal of Science* also published an article by William Hamilton Merritt whose account of the Welland Canal expounded on the scientific observations made at the Canadian canal site. See Spanagel, *De Witt Clinton and Amos Eaton*, 132.

140. See for example "Shipboard and Shoreline Science on Lake Ontario, 2008," COSEE Great Lakes Centers for Ocean Sciences Education Excellence. http://

coseegreatlakes.net/weblog. See also "Artist Paddles the Erie Canal from Buffalo to New York City." http://news.wbfo.org/post/artist-paddles-erie-canal-buffalo-nyc.

141. "Advertisement 1—no Title," *New York Farmer (1828–1837)* 10, no. 11 (July 30, 1837): 256.

Chapter 6

1. In this chapter, I analyze the connection between the canal age and tourism at Niagara and how canals and other internal improvements gave new shape and meaning to the northern tour as a distinctly transnational experience. In contrast, most studies of Niagara Falls focus on themes of national identify and pride, or on broader themes about nature and culture. Elizabeth McKinsey for example, argues that the power manifested in Niagara Falls becomes a metaphor for American prowess and patriotism. In *The End of America: The Beginning of Canada* (1988) Patrick McGreevy argues that the Niagara border has different meaning for the United States and Canada which can only be understood "in relation to two very different ideologies of nationalism." Ginger Strand likewise sees in Niagara's landscape a national icon "for both countries." McKinsey, *Niagara Falls*, 100; McGreevy, "The End of America: The Beginning of Canada," *The Canadian Geographer* 32, no. 4 (1988): 310; Strand, *Inventing Niagara*, 57. On the Canadian side, Patricia Jasen analyzes the Falls in terms of Canadian nationalism and identity, stating "the falls were a symbol of the 'New World' rather than 'American nationalism.'" Jasen, "Romanticism, Modernity, and the Evolution of Tourism on the Niagara Frontier, 1790–1850," *Canadian Historical Review* 72, no. 3 (September 1991): 292. Linda Revie similarly comes to the subject of Niagara Falls "with a special sense of Canadian history." L. L. Revie, *The Wilds of Niagara: Constructions of the falls in fine arts, literary and scientific narratives, from discovery through the twentieth century* (PhD diss., Boston College, 1998), 18. While acknowledging some of the above themes, I focus more on the integrated New York and Welland canal systems in promoting cross-border tourism and development in this region, and the ways in which these man-made wonders became attractions in their own right.

2. William Irwin's *New Niagara* makes the point that travelers on their way to Niagara Falls enjoyed the natural landscape and countryside, but found New York's Erie Canal a "thrilling tourist attraction in and of itself." Another scholar writes that the Erie Canal generated "much literary writings" by tourists contemplating the wondrous scenery and natural landscape, but they also saw the canal "as a marvel, considered to be worthy of praise in its own right." This observation, though concerned solely with the Erie Canal, has helped focus my own view that tourists and visitors anticipated and reveled in both countries' transportation innovations as much as they did natural wonders like Niagara Falls. Irwin, *The New Niagara*,

14; Roger W. Hecht, *The Erie Canal Reader, 1790–1950* (Syracuse, NY: Syracuse University Press, 2003), 1.

3. John W. Barber, *Historical Collections of the State of New York* (New York: Clark, Austin and Company, 1851), 412–13.

4. Cornog, *The Birth of Empire*, 7.

5. Before the canal age, the northerly route from Albany to the Champlain and Montreal and back up the St. Lawrence to Lake Ontario and Niagara Falls was referred to by one Canadian newspaper as "a fashionable tour." *Upper Canada Herald*, November 9, 1819. For a description of the various routes on the northern tour see Reginald Stuart, *American Expansionism*, 152. For a more general discussion of Niagara Falls and the Northern Tour see Berton, *Niagara*, 66–74; McKinsey, *Niagara Falls*, 129–31; and Strand, *Inventing Niagara*, 203–04.

6. Koeppel, *Bond of Union*, 286; Bernstein, *Wedding of the Waters*, 332.

7. For a comparison of stage and canal travel see Taylor, *Transportation Revolution*, 142.

8. J. P. Merritt, *Biography*, 131.

9. George Seibel, *Bridges Over The Niagara Gorge* (Niagara Falls, Ont.: Niagara Falls Bridge Commission, 1991), 3.

10. *Rochester Telegraph*, July 27, 1824.

11. General Summary, *Christian Register (1821–1835)* 5 (July 8, 1826): 107.

12. Joseph Ingraham, *A Manual for the Use of Visitors to the Falls of* Niagara (Buffalo: Printed for the Author by Charles Faxon, 1834), 15.

13. Charles Augustus Murray, *Travels in North America during the years 1834, 1835 & 1836, including a summer residence with the Pawnee tribe of Indians in the remote prairies of the Missouri, and a visit to Cuba and the Azore Islands* (London, Richard Bentley, 1839), 79–81. Online Text. Retrieved from the Library of Congress, https://www.loc.gov/item/02000373/. (Accessed April 03, 2017).

14. *British American Journal*, March 18, 1834.

15. *The American Traveller: Or, Guide Through the United States* (Philadelphia, 1836), 87.

16. *Time Magazine*, April 20, 1925.

17. *Prominent features of a northern tour. Written from a brief diary, kept in travelling from Charleston, S. C. to, and through Rhode Island, Massachusetts, New-Hampshire, Vermont, Lower and Upper Canada, New York, Maine, North-Carolina, South Carolina, and back to Charleston again, Commencing on the 12th of June, 1821, and terminating the 12th of November following* (Charleston, Printed for the Author by C. C. Sebring, 1822), 21–22. Online Text. Retrieved from the Library of Congress, https://www.loc.gov/item/13014403/. (Accessed April 03, 2017).

18. Mary E. Dewey, editor, *Life and Letters of Catherine M. Sedgwick* (New York: Harper and Brothers Publishers, 1871), 125–27. The canal commission report was dated February 1820 in *Documentary Sketch of the New York Canals by S. H. Sweet, Deputy State Engineer and Surveyor. Accompanying State Engineer and Surveyor's Report for 1862* (Albany, NY: VanBenthuysen Printer, 1863), 111.

19. Hecht, *The Erie Canal Reader*, 87–88.

20. Captain Basil Hall, *Travels in North America*, 1:127.

21. William Lyon Mackenzie, *Sketches of Canada and the United States* (London, Effingham Wilson, 1833), 5. Online Text. Retrieved from the Library of Congress, https://www.loc.gov/item/02010564/, (accessed April 03, 2017); "Domestic." *Western Recorder* (1824–1833), April 24, 1827.

22. William Lyon Mackenzie, *Sketches of Canada and the United States* (London, Effingham Wilson, 1833), 5. Online Text. Retrieved from the Library of Congress, https://www.loc.gov/item/02010564/, (accessed April 3, 2017).

23. James Stuart, *Three Years in North America* (Edinburgh, Printed for Robert Cadell; etc., etc., 1833), 75. Online Text. Retrieved from the Library of Congress, https://www.loc.gov/item/02000391. (accessed April 03, 2017).

24. Hecht, *The Erie Canal Reader*, 78–79.

25. In most if not all of the North American travel literature, the Erie Canal is given sole credit for the post-War of 1812 tourist boon at Niagara Falls, whereas the Oswego Canal and even the Welland is largely ignored. This is evident in most works. See for example Berton, *Niagara*, 48; Donaldson, *Niagara!*, 142; Stuart mentions the Oswego Canal in his chapter on the Northern Tour but notes that it was a "less common route." Stuart, *United States Expansionism*, 152.

26. *Oswego Press*, June 19, 1833.

27. *Oswego Palladium*, July 17, 1833.

28. The *British Whig*, June 17, 1834.

29. D. Fulton, ed., *New York to Niagara, 1836: the journal of Thomas S. Woodcock* (New York, 1938).

30. Rev. Issac Fidler, *Observations on Professions, Literature, Manner and Emigration in the United States and Canada* (New York: J and J Harper, 1833), 149.

31. "Article 11—no Title," *Atkinson's Saturday Evening Post (1831–1839)* 14, no. 733 (August 15, 1835): 1.

32. Frederick Marryat, *A Diary in America; with Remarks on its Institutions* (New York: D. Appleton, 1839), 44. For more on the prosperity of the salt works in this region see Mark Kurlansky, *Salt: A World History* (New York, NY: Penguin Books, 2002), 245–49. For salt manufacturing and trade in the 1830s, see "Art. VII—Trade and Manufacturing of Salt in the United States," *The Merchants Magazine and Commercial Review* 8, no. 4 (April 1, 1843).

33. Robert J. Vandewater, *The Tourist, or Pocket Manual for Travellers on the Hudson River, the Western and northern canals and railroads; the stage routes to Niagara Falls; and down Lake Ontario and the St. Lawrence to Montreal and Quebec* (New York: Harper and Brothers, 1838), 43–45.

34. In contrast, Linda Revie writes that while travelers welcomed the advances made possible by roads, canals, etc., "they disapproved of the way nature had been ruined by developments." Revie, *The Wilds of Niagara*, 260. For salt duties see *Canal Laws*, 1:326; Nathan Miller, *The Enterprise of a Free People*, 146.

35. *The New York State Tourist: Descriptive of the Scenery of the Hudson, Mohawk, and St. Lawrence Rivers. Falls, Lakes, Mountains, Springs, Rail Roads, Canals* (New York: Published by A. T. Goodrich, 1842), 126.

36. John Disturnell, *A Traveller's Guide Through the State of New York, Canada, Etc,. Embracing a General Description of the City of New York; the Hudson River Guide, and the Fashionable Tour to the Springs and Niagara Falls; with the Steam-boat, Rail-Roads, and Stage Routes* (New York: J. Disturnell, 1836), 49.

37. "Article 11—no Title," *Atkinson's Saturday Evening Post (1831–1839)* 14, no. 733 (August 15, 1835): 1.

38. Despite some accounts that salmon had disappeared because of the canalizing of the Oswego River, travelers referred to seeing large salmon in the Oswego–Lake Ontario water system. For an 1825 account on the abundance of salmon in Oneida and Onondaga Lakes see *Wayne Sentinel*, June 8, 1825.

39. "Editorial Correspondence." *The New Yorker (1836–1841)* 7, no. 16 (July 6, 1839): 249.

40. Vandewater, *The Tourist, or Pocket Manual for Travellers on the Hudson River, the western and northern canals . . . to Niagara Falls . . . Lake Ontario and the St. Lawrence to Montreal and Quebec*, 44.

41. *Oswego* Palladium, May 26, 1830.

42. *Oswego Palladium*, May 19, 1838. In addition to their attraction as "something ingenious and new" writes scholar Frank Macey, steamboats also offered faster, more reliable service. Mackey, *Steamboat Connections*, ix. See also Taylor, *The Transportation Revolution*, 60.

43. *The New York State Tourist*, 125–26.

44. Thomas Fowler, *The Journal of a Tour Through British America to the falls of Niagara* (Aberdeen: Lewis Smith, 1832), 208, 260.

45. C. D. Arfwedson, Esq. *The United States and Canada in 1832, 1833, and 1834*, 2 vols. (London, 1834)2: 256.

46. This assurance was stated in the *Oswego Palladium*, May 27, 1835. Taylor writes that steamboat races were common during this time, and by 1860 "not infrequently achieved better than twenty miles an hour." Taylor, *The Transportation Revolution*," 60.

47. *Kingston Chronicle*, July 2, 1831; *Kingston Chronicle and Gazette*, November 3, 1841. See also "Captain Joseph Whitney," *Stories* (November 2, 2016) in Maritime History of the Great Lakes digital library, http://stories.maritimehistoryofthegreatlakes.ca/captain-joseph-whitney/.

48. S. A. Ferrall, *A Ramble of Six Thousand Miles through the United States of America* (London: Published by Effingham Wilson, 1832), 21. Online Text. Retrieved from the Library of Congress, https://www.loc.gov/item/02000376/, (accessed April 3, 2017).

49. J. Benwell, *An Englishman's Travels in America: his observations of life and manners in the free and slave states* (London: Binns and Goodwin, 1853), 39.

50. James Stuart, *Three Years in North America*, 55. There are few book length studies on the Champlain Canal. However Thomas X. Grasso briefly discusses this canal and its ties to Canada in Grasso, "Champlain Canal," *Encyclopedia of New York State*, ed. Peter R. Eisenstadt and Laura-Eve Moss (Syracuse NY: Syracuse University Press, 2005), 304. For more on the history of commercial and recreational ties between the Champlain Valley and Lower Canada see Stuart, *United States Expansionism*, 48–49 and 67–68 and 115 and 152.

51. C. P. Lucas, *Lord Durham's Report, on the Affairs of British North America*, 2:217. See also chap. 6, n.53.

52. Hansen, *The Mingling*, 1:110–11.

53. *Canadian Emigrant and Western District Advertiser*, December 1, 1831. In the 1820s and 1830s, Upper Canada's population swelled, with thousands of new settlers coming from the British Isles. As discussed earlier, many of these immigrants found their way to Upper Canada through both the Erie and Oswego Canals illuminating the permeability of the international border in upstate New York and Upper Canada. For European immigration to Canada see Meinig, *The Shaping of America*, 2:51. Nora Faires also analyzes migration across the border in "Leaving the 'Land of the Second Chance,'" in Bukowczyk et al., *Permeable Border*, esp., 80–81.

54. James Stuart, *Three Years in North America*, 77.

55. Quoted on the anniversary of Brock's death at Queenston Heights, *Upper Canada Gazette*, October 21, 1824.

56. *An Excursion through the United States and Canada during the years 1822–1823 By an English Gentlemen* (London: Baldwin, Cradock, and Joy, 1824), 394.

57. The topic of tea and other goods being smuggled into Upper Canada from the United States is reported in *The Journals of the Legislative Assembly of Upper Canada for the Years 1792–1824* (Toronto, Ont.: Printed and published by L. K. Cameron, etc., 1911–1915), 22. First quote comes from David R. Moore, *Canada and the United States, 1815–1830* (PhD diss., University of Chicago, 1910), 121–22.

58. Koeppel details the building of the aqueduct in *Bond of Union*, 326–30.

59. Allen G. Noble, "The Tale of a Trail: Material Culture Along the Ridge Road," *Voices: The Journal of New York Folklore* 27 (Spring–Summer, 2001): 39.

60. McGreevy, *Stairway to Empire*, 67.

61. *Prominent Features of a Northern Tour* (Charleston: Printed for the author by C. C. Sebring, 1822), 18.

62. "General Sketches—western New York," *The New York Farmer* (1831–1839) 7 (April 15, 1837): 118.

63. "Original Communications," *The New York Mirror: A Weekly Gazette of Literature and the Fine Arts (1823–1842)* 14, no. 5 (July 30, 1836): 36.

64. Like his brother Peter Buell Porter, Augustus was a well-respected citizen of the Niagara borderland. For the few accounts on Augustus see Victor Hugo Paltsits, "Judge Augustus Porter, Pioneer of Niagara Falls," *New York History* 18, no. 2 (April 1937): 136–51. See also *Narrative of Early Years of Judge Augustus Porter,*

written by himself in 1848, Buffalo and Erie County Historical Society *Publications* (Buffalo, NY, 1904) 7: 277–322; Albert H. Porter, *Reminiscences of Niagara From 1806 to 1872* (Niagara Falls, NY: William Pool Printer, 1872), 4–10; Charles Robinson, "Life of Augustus Porter," Buffalo and Erie County Historical Society *Publications* (Buffalo, NY, 1904) 7: 229–75.

65. Forsyth went on to note that "there was also a foundry at Lower Black Rock." However, his recommendation was refused because it would compete with other ferry services on the Canadian side. Quote taken from Seibel, *The Niagara Portage Road*, 126. See also Robert L. Fraser, "FORSYTH, WILLIAM (1771–1841)," in *Dictionary of Canadian Biography*, vol. 7, University of Toronto/Université Laval, 2003– , accessed May 5, 2017, http://www.biographi.ca/en/bio/forsyth_william_1771_1841_7E.html.

66. William Pool, editor. *Landmarks of Niagara County, New York* (D. Mason and Company, Publishers, 1897), 405–08; Theodore Vinal, *Niagara Portage: From Past to Present* (Buffalo, NY: Foster and Stewart, 1949), 52–54; Seibel, *The Niagara Portage Road*, 114.

67. Captain Basil Hall, *Travels in North America in the years 1827 and 1828*, 1:190–91.

68. The reference to the Boston investors in Canada comes from a Niagara visitor in 1829 but I can find no other information to verify this point. "Article 6—no Title," *Saturday Evening Post (1821–1830)* III, no. 428 (October 10, 1829): 2: See also description of Biddle's stairway by Douglas Farley, Erie Canal Discover Center Director, Lockport, NY, http://www.niagara2008.com/history35.html; Berton, *Niagara*, 72. Seibel, *The Niagara Portage Road*, 68–69.

69. The Porter brothers were closely involved in the construction of Biddle's stairway. Roger Whitman, Scott Eberle (editor) and David A. Gerber (editor), *The Rise and Fall of a Frontier Entrepreneur: Benjamin Rathbun,"Master Builder and Architect"* (Syracuse, NY: Syracuse University Press, 1996), 62. For a general account of the Porter brothers as tourist developers see Strand, *Inventing Niagara*, 69–75. Construction on Biddle's stairway began a month after the *Michigan's* launching.

70. Strand, *Inventing Niagara*, 65.

71. Mc Kinsey, *Niagara Falls*, 127 and 148. William Irwin also sees in the *Michigan* "the duel at the Falls between technology and nature" in *The New Niagara*, 24. For a more popular account of the *Michigan* see Berton, *Niagara*, 55–63.

72. *Niles Weekly Register*, September 22, 1827.

73. The *New York Observer*, reprinted in *Niles Weekly Register*, September 22, 1827.

74. "Article 2—No Title," *The New England Galaxy and United States Literary Advertiser* 10, no. 515 (September 21, 1827): 3.

75. Colden in his *Memoir* of the Erie Canal celebration writes "The public papers apprize us, that . . . a vessel called the Noah's Arc, . . . will bring, it is said,

to our metropolis, specimens of all manner of living things, to be found in the forests that surround the Falls of Niagara." Cadwallader Colden, *Memoir Prepared at the Request of the Committee of the Common Council of the City of New York at the Celebration of the Completion of the New York Canals* (New York, 1825), 88.

76. Cadwallader D. Colden, *Memoir*, 88.

77. A copy of this ad may be read in the *Black Rock Gazette*, August 4, 1827.

78. A. Porter from William Bird, Black Rock, July 25, 1827. Augustus Porter Papers, Item # 300.001–300.009. The Buffalo History Museum.

79. Porter also left his closest friend property in the form of a village lot. Last will and testament of Peter B. Porter can be read in Peter B. Porter Papers, Roll 11, Misc. The Buffalo History Museum.

80. This term was used in the *Michigan* broadside in 1827 and appeared in Colden's description of the Erie Canal celebration two years earlier. Broadside can be viewed at The Buffalo History Museum, Buffalo New York.

81. *Black Rock Gazette*, August 4, 1827.

82. *Colonial Advocate*, May 24, 1827.

83. *Black Rock Gazette*, August 4 and September 22, 1827.

84. *Gore Gazette* reprinted in the *Montreal Gazette*, September 27, 1827. Courtesy of the Maritime History of the Great Lakes digital library.

85. *Black Rock Gazette*, June 22 and August 31, 1826; August 11, 1827. See also "Anecdotes of American Horses," *American Turf Register and Sporting Magazine (1829–1844)* 13 (October, 1842): 552. For more on this see Kevin J. Crisman and Arthur B. Cohn, *When Horses Walked On Water: Horse-Powered Ferries in Nineteenth-Century America* (Washington, DC: Smithsonian Institution Press, 1998).

86. See the advertisement in *Black Rock Gazette*, August 4, 1827, and George Seibel, *The Niagara Portage Road*, 122–25.

87. *The Ariel. A Literary Gazette*, 1, n.12 (October 6, 1827): 92.

88. *Gore Gazette*, September 8, 1827.

89. Augustus Porter to Peter B. Porter, Niagara Falls, August 29, 1827. Peter B. Porter Papers, Roll 11, Misc. The Buffalo History Museum.

90. *The Ariel. A Literary Gazette*, 1, n.12 (October 6, 1827): 92.

91. *Colonial Advocate*, September 13, 1827.

92. *Niles Weekly Register*, September 22, 1827.

93. Robert Bingham, *The Cradle of the Queen City*, Buffalo and Erie County Historical Society *Publications* (Buffalo, NY: 1931) 31: 464; "Article 5—"No Title," *Masonic Mirror: And Mechanics' Intelligencer (1824–1827)* 3, no. 33 (August 11, 1827): 253.

94. As quoted in the *Geneva Palladium*, September 26, 1827.

95. Patrick McGreevy writes that "as early as the 1830s, death had become part of the lure of Niagara." McGreevy, *Imagining Niagara*, 42.

96. *Colonial Advocate*, September 13, 1827.

97. *Niles Weekly Register*, September 22, 1827.

98. Crisfield Johnson, *Centennial History of Erie County, New York: Being its annals from the earliest recorded events to the hundredth year of American independence* (Buffalo, NY: Matthews and Warren, 1876), 380–82.

99. *Colonial Advocate*, September 13, 1827.

100. Elizabeth McKinsey writes that "most tourists seemed to revel" in such spectacles but the cruelty of sending defenseless animals to their inevitable death raised the ire of more than one spectator. With disgust did the *New York Spectator* comment on the "cruel and unusual" punishment that the animals were about to undergo, and another sickened observer wrote of "the useless sacrifice of animals for the gratification of a vicious curiosity" that was no different to "the Spanish custom of 'Bull fighting.'" First quote from McKinsey, *Niagara Falls*, 149; *New York Spectator*, August 10, 1827; "Article 5—no Title," *Masonic Mirror: And Mechanics' Intelligencer (1824–1827)* 33, no. 33 (August 11, 1827): 253.

101. It has been suggested by Donald Braider that Sam Patch got the idea to jump over the Falls in 1829 after witnessing the earlier releasing of the *Michigan*. See Donald Braider, *The Niagara* (New York: Holt, Rinehart, and Winston, 1972), 248–49.

102. See Buffalo Eagle Tavern Guest Book, The Buffalo History Museum.

103. Landon, *Western Ontario and the American Frontier*, 146; Gordon Donaldson, *Niagara!*, 121; Seibel, *The Niagara Portage Road*, 84–85.

104. *Farmer's Journal and Welland Canal Intelligencer*, October 10, 1827.

105. *Niles Weekly Register*, September 22, 1827.

106. *Black Rock Gazette*, August 1827, as quoted in the *Farmer's Journal and Welland Canal Intelligencer*, August 29, 1827. For reference to ferry as "curiosity" see John Fowler, *Journal of a Tour in the State of New York, in the Year 1830* (London: Whittaker, Treacher, and Arnot, 1831), 107 and 133.

107. "Genesee Falls—Opposition Rage among the Packet Boats—Lockport—Buffalo—its Enterprise," *Zion's Herald (1823–1841)* 7, no. 36 (September 7, 1836): 142.

108. "Retrospect of Western Travel," *Brown's Literary Omnibus; News, Books Entire, Sketches, Reviews, Tales, Miscellaneous Intelligence (1838–1838)* 2, no. 14 (April 6, 1838): 6.

109. Irwin, *The New Niagara*, 14; McGreevy, *Stairway to Empire*, 99.

110. Lawson, editor, *Nathan Roberts*, 36.

111. "A Tour," *The Western Monthly Review (1827–1830)* 2, no. 5 (October, 1828): 249.

112. Quoted in Sheriff, *The Artificial River*, 31.

113. James Boardman, *America, and the Americans*. (London: Printed for Longman, Rees, Orme, Brown, Green, & Longman, 1833), 132. Online Text. Retrieved from the Library of Congress, https://www.loc.gov/item/01026745/, (accessed April 4, 2017).

114. D. Fulton, ed., *New York to Niagara, 1836: The journal of Thomas S. Woodcock* (New York, 1938).

115. Gideon M. Davidson, *The Traveller's Guide Through the Middle and Northern States, and the Provinces of Canada* (Saratoga Springs, NY: G. M. Davidson, 1837), 254–55.

116. William Lyon Mackenzie, *Sketches of Canada and the United States*, 145.

117. "Editors Correspondence," *Christian Secretary (1822–1889)* 11, no. 20 (June 2, 1832): 78.

118. Hawthorne's visit to Niagara is recounted in Dow, *Anthology and Bibliography of Niagara Falls*, 1:194. See also Irwin, *The New Niagara*, 14.

119. Charles Lyell, *Travels in North America, Canada and Nova Scotia with Geological Observations* (London: J. Murray, 1845), 18.

120. Theodore Dwight, *The Northern Traveller; and Northern Tour: With the routes to the Springs, Niagara, and Quebec* (New York: J & J Harper, 1830), 92–93.

121. Captain Basil Hall, *Travels in North America*, 1:214.

122. *Third Report*, 221.

123. *Third Report*, 52 and 75.

124. Art. XXII—"Account of the Welland Canal, Upper Canada," *American Journal of Science and Arts (1820–1879)* 14, no. 1 (July, 1828): 159.

125. Joseph Pickering, *Emigration, or no emigration; being the narrative of the author, an English farmer from the year 1824 to 1830* . . . (London, Printed for the Author; Published by Longman, Rees, Orme, Brown, and Green, 1830), 91–92. Online Text. Retrieved from the Library of Congress, https://www.loc.gov/item/17013760/ (accessed April 4, 2017).

126. "Domestic." *Western Recorder (1824–1833)*, July 17, 1827. See also *Buffalo Emporium*, September 24, 1827 and *Black Rock Gazette*, September 29, 1827.

127. *Farmer's Journal and Welland Canal Intelligencer*, January 5, 1827.

128. "Article 1—no Title," *The Banner of the Constitution. Devoted to General Politics, Political Economy, State Papers, Foreign and Domestic News, &c. (1829–1832)* 1, no. 4 (December 26, 1829): 32.

129. Dwight, *The Northern Traveller*, 96.

130. "Article 1—no Title," *Saturday Evening Post (1821–1830)* II, no. 373 (September 20, 1828): 1.

131. *Mitchell's Compendium of the internal improvements of the United States . . . with a brief notice of . . . works of internal improvement in Canada and Nova Scotia* (Philadelphia: Mitchell and Hinman, 1835), 78. *The American Traveller*, 86. McKinsey interprets the Lockport Locks in *Niagara Falls*, 135.

132. *Table Rock Album and Sketches of the Falls and Scenery Adjacent* (Buffalo: Franklin Steam Printing House, 1862), 28.

133. Capt. Basil Hall, *Travels in North America*, 1: 208.

134. Godfrey T. Vigne, *Six Months in America* (Philadelphia: T. T. Ash, 1833), 154. The same letter can be found in "Article 1—no Title," *Waldie's Select Circulating*

Library (1832–1842) 1, no. 7 (Nov 28, 1832): 97. Discussing the tourist industry in the late twentieth century, John Jackson writes that the "Frontier's setting on two of the Great Lakes and its long history of canal and harbor development," has resulted in the existence of "a strong element of transboundary sharing" between the two nations. Jackson, *The Mighty Niagara*, 384 and 396.

Chapter 7

1. *Table Rock Album and Sketches of the Falls and Scenery Adjacent*, 42.
2. The Ericsson propeller was located at the stern of the ship and proved to be much more efficient than the cumbersome side-paddle wheels.
3. *Buffalo Commercial Advertiser and Journal*, April 2, 1841. See also *Vandalia* (Propeller), November 1, 1841. Maritime History of the Great Lakes digital library.
4. *Kingston Chronicle and Gazette*, November 20 and December 4, 1841; *Oswego Press*, November 1, 1841.
5. *St. Catharines Journal*, November 25, 1841; "Article 8—no title," *Scientific American* (1845–1908) XIV(November 5, 1881): 292.
6. *St. Catharines Journal*, November 25, 1841.
7. George Rawlyk, "Thomas Coltrin Keefer and the St. Lawrence–Great Lakes Commercial system," *Inland Seas* 19 (Fall 1963): 190; H. V. Nelles, "KEEFER, THOMAS COLTRIN," in *Dictionary of Canadian Biography*, vol. 14, University of Toronto/Université Laval, 2003– , accessed April 13, 2017, http://www.biographi.ca/en/bio/keefer_thomas_coltrin_14E.html.
8. *Papers Respecting Claim of Shareholders in the Late Welland Canal Company. Printed by Order of the Legislative Assembly* (Quebec, Ont.: Printed in Rollo Campbell, 1853), 4. See also Aitken, *The Welland Canal Company*, 100; Creighton, *The Empire of the St. Lawrence*, 270.
9. *Papers Respecting Claim of Shareholders in the Late Welland Canal Company. Printed by Order of the Legislative Assembly* (Quebec, Ont.: Printed in Rollo Campbell, 1853), 3–4.
10. J. P. Merritt, *Biography*, 162.
11. *Kingston Chronicle and Gazette*, January 24 and November 7, 1835.
12. Oswego Canal agent John Edwards for example placed ads in both the *Kingston Chronicle* and *Montreal Gazette* for several consecutive weeks, urging Canadian laborers to apply for work at Oswego. *Kingston Chronicle and Gazette*, May 7, 1836. Earlier quote regarding "a brisk trade" with Buffalo in *Hamilton Gazette*, reprinted in the *St. Catharines Journal*, June 1, 1837.
13. "Art. III—Lake Navigation of North America," *The Merchants' Magazine and Commercial Review (1839–1870)* 3, no. 3 (September 1, 1840): 216.

14. *St. Catharines Journal*, June 2, 1842. The first quote from the *New York Commercial Advertiser*, after applauding developments on the Canadian canal system continued: "Our neighbors deserve credit for what they are doing; but let us not sleep while they are so wide awake." New York *Commercial Advertiser* reprinted in *Kingston Chronicle and Gazette*, December 22, 1841. For the point that the enlargement of the Welland greatly benefited the Oswego trade see John C. Churchill, ed., *Landmarks of Oswego County, New York*, 349, and Crisfield Johnson, *History of Oswego, New York*, 74. William Kingsford also acknowledged that Oswego "has much to hope for the enlargement of the Welland Canal." William Kingsford, *The Canadian Canals: Their history and cost, with an inquiry into the policy necessary to advance the wellbeing of the province* (Toronto, Ont.: Rollo & Adam, 1865), 98.

15. Way, *Common Labour*, 237.

16. After a lifetime of supporting internal improvements and freer trade along the United States–Canada borderland, William Hamilton Merritt fittingly died aboard a steamer while passing through the Cornwall Canal on the St. Lawrence River. J. J. Talman, "MERRITT, WILLIAM HAMILTON (1793–1862)," in *Dictionary of Canadian Biography*, vol. 9, University of Toronto/Université Laval, 2003– , accessed April 13, 2017. http://www.biographi.ca/en/bio/merritt_william_hamilton_1793_1862_9E.html.

17. "Cargo barges traffic increases on New York canal system," May 22, 2014, at www.times.cunion.com. See also "20-year high expected for commercial traffic on canals" http://innovationtrail.org/post/20-year-high-expected-commercial-traffic-canals. A year earlier, NPR announced that commercial shipping on the Erie Canal "from Canada is expected to lead to a level of commercial traffic not seen in decades." http://www.npr.org/2013/06/25/195426326/commercial-shipping-revived-along-erie-canal.

18. William Gooding to William Hamilton Merritt, Lockport, 1853. Merritt Family Papers [microform], MG 24 E1, Brock University Archives; William Gooding, Chief Engineer, I. & M. Canal, no. 5 (Lockport, IL: Illinois Canal Society, 1982); McGreevy, *Stairway to Empire*, 130–32.

Bibliography

Primary Sources

MANUSCRIPTS

New York
Augustus Porter Papers. The Buffalo History Museum, Buffalo, New York.
Peter. B. Porter Papers. The Buffalo History Museum, Buffalo, New York.
Peter A. Porter. "The Niagara Ship Canal," unpublished papers and miscellaneous clippings.

Ontario, Canada
Merritt Family Papers [microform], MG 24 E1, Brock University Archives, St. Catharines, Ont.
Reports of the Welland Canal Company. Brock University Archives, St. Catharines, Ont.
Diary of Anne Powell on her voyage from Montreal to Detroit with her brother W. D. [William Dummer] Powell (later Chief Justice of Upper Canada). Jarvis family fonds, 1789–1847, n.d. RG 563, Brock University Archives, St. Catharines, Ont.

SELECT PUBLICATIONS OF THE BUFFALO HISTORY MUSEUM, BUFFALO, NEW YORK

The Holland Land Company and Canal Construction in Western New York: Buffalo–Black-Rock Harbor Papers, Journals and Documents. Buffalo, NY, 1910.
Reports of Joseph Ellicott. Edited by Robert W. Bingham. Buffalo, NY, 1937–1941.
Extracts from Joseph Ellicott Letter Books and Early Correspondence. Buffalo, NY, 1922.
Facts and Observations in Relation to the Origin and Completion of the Erie Canal. New York: H. B. Holmes, 1825.

NEW YORK STATE LEGISLATIVE RECORDS

Laws of the State of New York Passed at the Sessions of the Legislature Held in the Years 1797, 1798, 1799, and 1800. Albany, New York: Weed, Parson, and Company Printers 1887. 21st Session, 5th of April, 1789.
Laws of the State of New York in Relation to the Erie and Champlain Canals, Together with the Annual Reports of the Canal Commissioners and other Documents. 2 Vols. Albany, NY, 1825.
Journal of the Assembly of the State of New York.
Journal of the Senate of the State of New York.
Documentary Sketch of the New York State Canals by S. H. Sweet, Deputy State Engineer and Surveyor. Documents of the State of New York, 86th Session—1863, Vol. IV—Nos. 50–73 inclusive.

UNITED STATES GOVERNMENT DOCUMENTS

American State Papers. Documents, Legislative and Executive of the Congress of the United States, . . . March 3, 1789, and ending March 3, 1809. Washington, DC: Gales and Seaton, 1834.
American State Papers. "Report of Mr. Weston to the Directors of the Western and Northern Inland Lock Navigation Companies," Albany, NY, December 23, 1795. Vol. 1, Misc., 775.
Report of the Secretary of the Treasury; on the Subject of Public Roads and Canals . . . March 2nd, 1807. Washington, DC: William A. Davis, 1816.

CANADIAN GOVERNMENT DOCUMENTS

Bill to Approve and Amend the Communication between the Lakes Erie and Ontario, by Land and Water. Niagara: Printed by S and G Tiffany, Printers to the Provinces, 1799.
The Journals of the Legislative Assembly of Upper Canada for the Years 1792–1824. Toronto: Printed and Published by L. K. Cameron, etc., 1911–1915.
Appendix to the Journal of the House of Assembly of Upper Canada, Session 1836–37, Vol. 2. Titled No. 90. *Third Report From the Select Committee.* Toronto, Ont., 1836.
Report of the Committee of the Commons House of Assembly of Upper Canada, Relative to the Welland Canal, with the Letter of John B. Yates, Esq. St. Catharines, Ont., 1834.
Papers Respecting Claim of Shareholders in the Late Welland Canal Company. Printed by Order of the Legislative Assembly. Quebec: Printed in Rollo Campbell, 1853.

AMERICAN NEWSPAPERS

Black Rock Advocate
Black Rock Gazette
Black Rock Beacon
Buffalo Gazette
Buffalo Emporium
Buffalo Journal
Buffalo Republican
Free Oswego Press
Onondaga Register
Onondaga Register and Syracuse Gazette
Oswego County Whig
Oswego Palladium
Oswego Palladium and Chronicle
New York Albion
New York Spectator
Niles Weekly Register
Rochester Telegraph
Salina Sentinel

CANADIAN NEWSPAPERS

British Whig
Canada Constellation
Colonial Advocate
Farmer's Journal and Welland Canal Intelligencer
Kingston Chronicle
Kingston Chronicle and Gazette
Kingston Gazette
Niagara Gleaner
Niagara Herald
Niagara Reporter
St. Catharines Journal
Upper Canada Gazette
Upper Canada Herald
The Women's Standard

PRINTED SOURCES

"A Niagara Falls Tourist of the Year 1817 Being a Journal of Captain Richard Langslow." Buffalo Historical Society *Publications*, Vol. 5 (Buffalo, NY, 1902), 111–33.

"Account of the Welland Canal by William Hamilton Merritt, Esq." *American Journal of Science and Arts*, Vol. 14, No. 1 (July 1828).

American Tract Magazine for the Year 1828. Vol. 3, No. 10 (October, 1828). New York: D. Fenshaw Printer, 1828.

A Ride to Niagara in 1809 by T. C. Rochester, New York, 1915.

An Excursion through the United States and Canada during the Years 1822–1823. By an English Gentleman. London: Baldwin, Cradock, and Joy, 1824.

"An Itinerary to Niagara Falls in 1809." *Pennsylvania Magazine of History and Biography*, Vol. XXIV. 1900.

Arfwedson, C. D. Esq. *The United States and Canada in 1832, 1833, and 1834*. 2 Vols. London: Richard Bentley, New Burlington Street, 1834.

A Serious Appeal to the Wisdom and Patriotism of the Legislature of the State of New York; on the subject of a Canal Communication between the Great Western Lakes. By a Friend to His Country. New York, 1816.

Barber, John W. *Historical Collections of the State of New York*. New York: S. Tuttle, 1845.

Benwell, J. *An Englishman's Travels in America: His observations of life and manners in the free and slave states*. London: Ward and Lock, 1857.

Blane, William N. *Travels Through the United States and Canada*. London: Baldwin and Co., 1828.

Boardman, James. *America and the Americans, By a Citizen of the World*. London, 1833.

Bronson, Alvin. *An essay on the commerce and transportation of the vallies [sic] of the Great Lakes and rivers, of the North-west*. Oswego, NY; Advertiser and Times Steam Press, 1868.

Bullock, William. *Bullock's Journey from New York to New Orleans in 1827*. London: John Miller, 1827.

Burr, David H. *An Atlas of the State of New York*. New York: Published by David H. Burr, 1829.

Callaghan, E. B. *The Documentary History of the State of New York*. 4 Vols. Albany, NY: Weed, Parsons, & Co., Public Printers, 1849.

Campbell, William W. *The Life and Writings of De Witt Clinton*. New York: Baker and Scribner, 1849.

The Canal Policy of the State of New York; Delineated in a letter to Robert Troup, Esquire. By Tacitus. Albany, NY, 1821.

Coke, E. T. *A Subaltern's Furlough: descriptive of scenes in various parts of the United States, upper and lower Canada, New-Brunswick, and Nova Scotia, during the summer and autumn of 1832*. New York: J. & J. Harper, 1833.

Colden, Cadwallader D. *Memoir Prepared at the Request of a Committee of the Common Council of the City of New York, and Presented to the Mayor of the City, at the Celebration of the Completion of the New York Canals*. New York, 1825.

Creighton, Ogden. *General View of the Welland Canal in the Province of Upper Canada.* London, 1830.

Darby, William, and Theodore Dwight, Jr. *A New Gazetteer of the United States of America: Containing a Copious Description of the States, Territories, Counties . . . Including Other Interesting and Valuable Information.* Hartford: Published by Edward Hopkins, 1833.

"Description of the River and Falls of Niagara, and the Country Bordering Upon the Navigable Part of the River Below the Falls." Article 1—No Title: From Weld's Travels. *The Literary Magazine, and American Register* (1803–1807), Vol. 2, No. 12 (September 1804).

Disturnell, John. *A Traveller's Guide Through the State of New York, Canada, Etc. Embracing a General Description of the City of New York; the Hudson River Guide, and the Fashionable Tour to the Springs and Niagara Falls; with the Steam-boat, Rail-Roads, and Stage Routes.* New York: Published by John Disturnell, 1836.

Dwight, Theodore. *The Northern Traveller; and Northern Tour: With the routes to the Springs, Niagara, and Quebec.* New York: J & J Harper, 1830.

Ferrall, S. A. *A Ramble of Six Thousand Miles Through the Untied States of America.* London: Published by Effingham Wilson, 1832.

Fowler, Thomas. *The Journal of a Tour Through British America to the Falls of Niagara.* Aberdeen: Lewis Smith, 1832.

Galt, John. *The Autobiography of John Galt in Two Volumes.* Philadelphia: Key and Biddle, 23 Minor Street, 1834.

Geddes, George. *Origin and History of the Measures that Led to the Construction of the Erie Canal.* Syracuse, NY: Simmers and Company, 1866.

Godfrey, T. Vigne. *Six Months in America.* Philadelphia: T. T. Ash, 1833.

Granger, Gideon, Esq. *Speech Delivered before a Convention of the People of Ontario County, New York . . . on the Subject of a Canal from Lake Erie to the Hudson.* Canandaigua, NY, 1817.

Haines, Charles. *Considerations on the Great Western Canal From the Hudson to Lake Erie.* New York: Spooner and Worthington Printers, 1818.

Hall, Captain Basil. *Travels in North America in the Years 1827 and 1828.* London: Simpkin and Marshall, 1830.

Hamilton, Thomas. *Men and Manners in America.* Philadelphia, PA: Carey, Lea and Blanchard, 1833.

Hennepin, Louis. *A New Discovery of a Vast Country in America.* London: Printed for Henry Bonwicke, 1699.

Heriot, George. *Travels Through the Canadas.* London, 1807.

Hodgson, Adam. *Letters from North America, Written During a Tour in the United States and Canada.* 2 Vols. London, 1824.

Hosack, David. *Memoir of De Witt Clinton.* New York: J. Seymour, 1829.

Howison, John. *Sketches of Upper Canada.* Edinburgh: Oliver and Boyd Publishers, 1822.

Ingraham, Joseph. *A Manual for the Use of Visitors to the Falls of Niagara.* Buffalo, NY: Printed for the Author by Charles Faxon, 1834.

In Memoriam: Orson Phelps, Esq. Portland, OR: Geo. H. Himes, Printer, 1870.

Izard, Ralph. *An Account of a Journey to Niagara, Montreal and Quebec in 1765.* New York: Printed by Wilson Osborn, 1846.

Letters on the Natural History and Internal Resources of the State of New York. By Hibernicus [De Witt Clinton] Author. New York, 1822.

Lyell, Charles. *Travels in North America, Canada and Nova Scotia with Geological Observations.* 2 Vols. London: J. Murray, 1845.

Marryat, Captain Frederic. *Diary in America.* New York: D. Appleton, 1839.

Maude, John. *Visit to the Falls of Niagara in 1800.* London: Longman's, Rees, Orme, Brown & Green, 1826.

Mackenzie, William Lyon. *Sketches of Canada and the United States.* London: Effingham Wilson, 1833.

Melish, John. *Travels Through the United States of America in 1806 and 1807, and 1809, 1810, and 1811.* Philadelphia, PA: Printed for the Author, 1818.

Merritt, William Hamilton. *Brief Review of the Origin, Progress, Present State, and Future Prospects of the Welland Canal.* St. Catharines: H. Leavenworth, 1852.

Merritt, J. P. *Biography of the Hon. W. H. Merritt, MP.* St. Catharines: E. S. Leavenworth Est., 1875.

Mitchell's Compendium of the Internal Improvements of the United States . . . with a brief notice of works of internal improvements in Canada and Nova Scotia. Philadelphia, PA: Mitchell and Hinman, 1835.

Murray, Charles Augustus. *Travels in North America during the years 1834, 1835, and 1836.* 2 Vols. London: Richard Bentley, New Burlington Street, 1839.

A Narrative of the Life, Travels, and Adventures of Captain Israel Adams: Who Lived at Liverpool, Onondaga County, New York. New York: D. Bennett, 1848.

Pickering, Joseph. *Inquiries of an emigrant; being a narrative of an English Farmer from the year 1824–1830; during which time he traversed the United States of America and the British Province of Canada.* London: E. Wilson, 1830.

Ploughshare, Peter. *Considerations Against Continuing The Great Canal West of Seneca; Addressed to the Members Elect of the Legislature of the State of New York.* Utica, New York: Printed by William Williams, 1819.

Records of the Village of Oswego, 1828–1840. Oswego, NY: Oswego, Daily and Weekly Times Print, 1874.

Rochefoucauld-Liancourt, F. A. *Travels through the United States of North America . . . in the years 1795, 1796, and 1797.* London: Printed for R. Philips, 1799.

Spafford, Horatio Gates. *A Gazetteer of the State of New York: Embracing an Ample Survey and Description of its Counties, Towns, Villages, Canals. . . .* Albany, NY, 1824.

Speeches of Gerrit Smith in Congress, 1853–1854. New York: Mason Brothers, 1855.

Stuart, James. *Three Years in North America.* 2 Vols. Edinburgh: Robert Cadell, 1833.

Table Rock Album and Sketches of the Falls and Scenery Adjacent. Author Unknown. Buffalo, NY, 1862.

Tatham, William. *The Political Economy of Inland Navigation.* London: Robert Faulder, 1799.

Todd, Henry Cook. *Notes Upon Canada and the United States from 1832 to 1840.* Toronto, Ont.: Printed by Rogers and Thompson, 1840.

Troup, Robert, Esq. *A Letter to the Hon. Brockholst Livingston, Esq., on the Lake Canal Policy of the State of New York.* Albany, NY, 1822.

Turner, Orsamus. *Pioneer History of the Holland Land Purchase of Western New York.* Buffalo, NY: Jewett, Thomas and Company, 1849.

Turner, Orasmus. *History of the Pioneer Settlement of Phelps and Gorham's Purchase and Morris' Reserve: Embracing the counties of Monroe, Ontario, Livingston, Yates, Steuben, most of Wayne and Allegany, and parts of Orleans, Genesee, and Wyoming: to which is added, a supplement or extension of the pioneer history.* Rochester, NY: W. Alling, 1851.

Vandewater, Robert J. *The Tourist, or Pocket Manual for travelers on the Hudson River, the Western and northern canals and railroads; the stage routes to Niagara Falls; and down Lake Ontario and the St. Lawrence to Montreal and Quebec.* 6th edition. New York: Harper and Brothers, 1838.

Watson, Elkanah. *History of the Rise, Progress, and Existing Condition of the Western Canals of the State of New York.* Albany, NY: Packard and Van Benthuysen Printers, 1820.

Watson, Winslow. *Men and Times of the Revolution: Or, Memoirs of Elkanah Watson, including Journals of Travels in Europe and America.* New York: Dana and Company Publishers, 1856.

Weld, Isaac. *Travels through the States of North America, and the Provinces of Upper and Lower Canada.* 2 Vols. London: Printed for John Stockdale, Piccadilly, 1799.

The Western Guidebook and Emigrant's History: Containing General Descriptions of Different Routes through the State of New York, Ohio, Indiana, Illinois, and Michigan. Buffalo, NY: Oliver G. Steele Publisher, 1834.

SECONDARY SOURCES

Aikins, Barton. *Modern Antiquities: Sketches of Early Buffalo and the Great Lakes.* Buffalo, NY, 1898.

Aitken, Hugh G. J. *The Welland Canal Company.* St. Catharines, Ont.: The Canadian Canal Society, 1997. First published 1954 by Harvard University Press.

Andreae, C. *Archeological Excavation of Lock 24, First Welland Canal.* London, Ont.: The Company, 1988.

Bangs, Jeremy Dupertuis, *The Travels of Elkanah Watson: An American Businessman in the Revolutionary War, in 1780s Europe and in the Formative Decades of the United States.* Jefferson, NC: McFarland & Company, Inc., Publishers, 2015.

Bannan, Theresa. *Pioneer Irish of Onondaga, about 1776–1847.* New York: G. P. Putnam's Sons, 1911.

Bernstein, Peter L. *Wedding of the Waters.* New York: W. W. Norton and Company, 2005.

Berton, Pierre. *Niagara: A History of the Falls.* Toronto, Ont.: McClelland and Stewart Inc., 1992.

Bothwell, Robert. *Your Country, My Country: A Unified History of the United States and Canada.* Oxford, UK: Oxford University Press, 2015.

Bowler, R. Arthur, ed. *War along the Niagara: Essays on the War of 1812 and its Aftermath.* New York: Old Fort Niagara Association, Inc., 1991.

Braider, Donald. *The Niagara.* New York: Holt, Rinehart, and Winston, 1972.

Brebner, John Bartlet. *North Atlantic Triangle: The Interplay of Canada, the United States and Great Britain.* New Haven, CT: Yale University Press, 1945.

Brooks, Charles E. *Frontier Settlement and Market Revolution: The Holland Land Purchase.* Ithaca, NY: Cornell University Press, 1996.

Bukowczyk, John, et al. *Permeable Border: The Great Lakes Basin as Transnational Region, 1650–1990.* Pittsburgh, PA: University of Pittsburgh Press, 2005.

Burroughs, Edwin, and Mike Wallace. *Gotham: A History of New York City to 1898.* Oxford, UK: Oxford University Press, 1999.

Campbell, Marjorie. *Niagara: Hinge of the Golden Arc.* Toronto, Ont.: Ryerson Press, 1958.

Carter, Goodrich. *Government Promotions of American Canals and Railroads, 1800–1890.* New York: Columbia University Press, 1960.

Cartwright, C. E. *Life and Letters of the Late Honourable Richard Cartwright.* Toronto, Ont.: Belford Brothers, 1876.

Chazanof, William. *Joseph Ellicott and the Holland Land Company.* Syracuse, NY: Syracuse University Press, 1970.

Condon, George. *Stars in the Water: The Story of the Erie Canal.* New York: Doubleday, 1974.

Cornog, Evan. *The Birth of Empire: De Witt Clinton and the American Experience, 1769–1828.* Oxford, UK: Oxford University Press, 1998.

Cowan, Helen L. *Charles Williamson: Genesee Promoter, Friend of Anglo-American Rapprochement.* Clifton, NJ: Augustus M. Kelley Publishers, 1941. Reprinted in 1973 by Augustus M.Kelley Publishers, Clifton, New Jersey.

Creighton, Donald. *The Empire of the St. Lawrence.* Toronto, Ont.: The MacMillan Company of Canada Limited, 1956.

Crisman, Kevin J., and Arthur B. Cohn. *When Horses Walked on Water: Horse-Powered Ferries in Nineteenth Century America.* Washington, DC: Smithsonian Institution Press, 1988.

Cross, Whitney R. *The Burned-Over District: The Social and Intellectual History of Enthusiastic Religion in Western New York, 1800–1850*. New York: Harper Torchbooks, 1950.

Cruickshank, E. *The Correspondence of Lieut. John Graves Simcoe, with allied documents relating to his administration of the government of Upper Canada*. 5 Vols. Toronto, Ont.: 1923–1931.

Dewey, Mary E. Editor. *Life and Letters of Catherine M. Sedgwick*. New York: Harper and Brothers Publishers, 1871.

Donaldson, Gordon. *Niagara! The Eternal Circus*. Toronto, Ont.: Doubleday Canada Ltd., 1979.

Dow, Charles Mason. *Anthology and Bibliography of Niagara Falls*. 2 Vols. Albany, NY: J. B. Lyon Company, Printers, 1921.

Doyle, James. *Yankees in Canada: A Collection of Nineteenth Century Travel Narratives*. Toronto, Ont.: ECW Press, 1980.

Drescher, Nuala. *Engineers for the Public Good: A History of the Buffalo District U.S. Army Corps of Engineers*. Buffalo, NY: The District, 1982.

Dunfield, R. W. *The Atlantic Salmon in the History of North America*. Ottawa, Canada: Department of Fisheries and Oceans, 1985.

Edmonds, Walter. *Rome Haul*. Syracuse NY: Syracuse University Press, 1929.

Edson, Obed. *Biographical and Portrait Cyclopedia of Chautauqua County*. Philadelphia, PA: Published by John M. Gresham and Co, 1891.

Errington, Jane. *The Lion, the Eagle, and Upper Canada*. Montreal & Kingston: McGill-Queen's University Press, 1987.

Fisher, Bruce. *Borderland: Essays from the US–Canadian Divide*. Albany, NY: SUNY Press, 2015.

Frothingham, Octavius B. *Gerrit Smith: A Biography*. New York: G. P. Putnam Sons, 1879.

Garrity, Richard. *Canal Boatman: My Life on Upstate Waterways*. New York: Syracuse, NY: Syracuse University Press, 1977.

Gerber, David A. *The Making of an American Pluralism: Buffalo, New York, 1825–1860*. Urbana and Chicago, IL: University of Illinois Press, 1989.

Glazebrook, G. P. de T. *A History of Transportation in Canada*. Toronto, Ont.: The Ryerson Press, 1938.

Goldman, Mark. *High Hopes: The Rise and Decline of Buffalo, New York*. Albany, NY: SUNY Press, 1983.

Hammond, Jabez. The History of Political Parties in the State of New York. Buffalo, NY: Phinney and Company, 1850.

Hansen, Marcus Lee. *The Mingling of the Canadian and American Peoples*. 2 Vols. New Haven, CT: Yale University Press, 1940.

Hecht, Roger W. *The Erie Canal Reader, 1790–1950*. Syracuse, NY: Syracuse University Press, 2003.

Holley, George W. *The Falls of Niagara and other Famous Cataracts*. New York: A. C. Armstrong and Son, 1883.

Horton, John T., et al. *History of Northwestern New York*. 3 Vols. New York: Lewis Historical Publishing Company, 1947.

Hotchkiss, William O. *Early Days of the Erie Canal*. Princeton, NJ: Princeton University Press, 1940.

Howells, W. D, Mark Twain, et al. *The Niagara Book*. New York: Doubleday, Page, and Company, 1901.

Hulbert, Archer Butler. *The Niagara River*. New York: Putnam, 1908.

Irwin, William. *The New Niagara: Tourism, Technology and the Landscape of Niagara Falls, 1776–1917*. University Park, PA: The Pennsylvania State University Press, 1996.

Jackson, John N. *The Mighty Niagara: One River—Two Frontiers*. Amherst, NY: Prometheus Books, 2003.

Jennings, Walter W. *The American Embargo 1807–1809: With particular reference to its effect on industry*. Iowa City, IA: The University, 1921.

Johnson, Crisfield. *History of Oswego County, New York*. New York: L. H. Everts and Company, 1877.

Johnson, Paul. *A Shopkeeper's Millennium: Society and Revivals in Rochester, New York, 1815–1837*. New York: Hill and Wang, 1978.

Jones, Robert L. *The History of Agriculture in Ontario, 1613–1880*. Toronto, Ont.: University of Toronto Press, 1946.

Karr, Clarence. *The Canada Land Company: The Early Years—An Experiment in Colonization, 1823–1843*. Toronto, Ont.: Ontario Historical Society, 1974.

Kelly, Jack. *Heaven's Ditch: God, Gold, and Murder on the Erie Canal*. New York: St. Martin's Press, 2016.

Kingsford, William. *The History of Canada*. 10 Vols. Toronto, Ont.: Boswell and Hutchinson, 1887.

Keefer, Thomas C. *The Old Welland Canal and the Man Who Made It*. St. Catharines, Ont.: The Print Shop, 1920.

Koeppel, Gerard. *Bond of Union: Building the Erie Canal and the American Empire*. Cambridge, MA: Da Capo Press, 2009.

Kurlansky, Mark. *Salt: A World History*. New York: Penguin Books, 2002.

Landes, David S., et al., eds. *The Invention of Enterprise: Entrepreneurship from Ancient Mesopotamia to Modern Times*. Princeton, NJ: Princeton University Press, 2012.

Landon, Fred. *Western Ontario and the American Frontier*. Toronto, Ont.: McClelland and Stewart Limited, 1967.

Larkin, Daniel F. *New York State Canals: A Short History*. New York: Purple Mountain Press, 1998.

Larson, John L. *Internal Improvement: National Public Works and the Promise of Popular Government in the Early United States*. Chapel Hill, NC: University of North Carolina Press, 2001.

Lawson, Eric. *Nathan Roberts: Erie Canal Engineer.* New York: North Country Books, Inc., 1997.
Lecker, Robert, ed. *Borderlands: Essays in Canadian-American Relations.* Toronto, Ont.: ECW Press, 1991.
Legget, Robert. *Rideau Waterway.* Toronto, Ont.: University of Toronto Press, 1955.
Lincoln, Charles. Z. *State of New York. Messages from the Governors, Comprising Executive Communications to the Legislature and other Papers Relating to Legislation from the Organization of the First Colonial Assembly in 1683 to and including the Year 1906.* New York: J. B. Lyon Publisher, 1909.
Lord, Philip, Jr. *The Navigators: A Journal of a Passage on the Inland Waterways of New York, 1793.* New York: New York State Museum, 2003.
Lower, A. R. M. *The North American Assault on the Canadian Forest: A History of the Lumber Trade between Canada and the United States.* New York: Greenwood Press, 1968.
Lucas, C. P. *Lord Durham's Report on the Affairs of British North America.* 3 Vols. Oxford, UK: Oxford at the Clarendon Press, 1912.
Mackey, Frank. *Steamboat Connections: Montreal to Upper Canada, 1816–1843.* Montreal and Kingston: McGill-Queen's University Press, 2000.
Mahon, John K. *The War of 1812.* Gainesville, FL: The University of Florida Press, 1972.
Marshall, Orasmus. *The Niagara Frontier: Embracing Sketches of its Early History, and Indian, French, and English Local Names.* Buffalo, NY: J. Warren and Company Publishers, 1865.
McCalla, Douglas. *Planting the Province: The Economic History of Upper Canada, 1784–1870.* Toronto, Ont.: University of Toronto Press, 1993.
McConnell, Michael N. *Army and Empire: British Soldiers on the American Frontier, 1758–1775.* Lincoln, NE: University of Nebraska Press, 2004.
McGreevy, Patrick. *Stairway to Empire: Lockport, the Erie Canal, and the Shaping of America.* Albany, NY: SUNY Press, 2009.
McGreevy, Patrick. *Imagining Niagara: The Meaning and Making of Niagara Falls.* Amherst, MA: University of Massachusetts Press, 1994.
McKinsey, Elizabeth. *Niagara Falls: Icon of the American Sublime.* Cambridge, UK: Cambridge University Press, 1985.
McMahon, Helen G. *Chautauqua County: A History.* Buffalo, NY: Henry Stewart Inc., 1958.
McNall, Neil Adams. *An Agricultural History of the Genesee Valley, 1790–1860.* Philadelphia, PA: University of Philadelphia Press, 1952.
Meinig, D. W. *The Shaping of America: A Geographical Perspective on 500 Years of History.* 2 Vols. New Haven, CT: Yale University Press, 1993.
Merritt, Richard, et.al. *The Capital Years: Niagara-on-the-Lake, 1792–1796.* Toronto, Ont.: Dundurn Press, 1991.

Miller, Nathan. *The Enterprise of a Free People: Prospects of Economic Development in New York State during the Canal Period, 1792–1838.* Ithaca, NY: Cornell University Press, 1962.

O'Callaghan, Edmund B. *The Documentary History of the State of New York.* 4 Vols. Albany, NY: Weed, Parsons and Company, Public Printers, 1849.

Passfield, Robert W. *Military Paternalism: Labour and the Rideau Canal Project.* Bloomington, IN: Author House, LLC, 2013.

Pool, Richard, ed. *Landmarks of Niagara County, New York.* New York: D. Mason & Company Publishers, 1897.

Porter, Albert H. *Reminiscences of Niagara, from 1806–1872.* Niagara Falls, NY: William Pool Publishers, 1872.

Randall, J., and Herman Konrad, *North America Without Borders: Integrating Canada, the United States, and Mexico.* Calgary, Alberta: University of Calgary Press, 1992.

Read, D. B. *The Life and Times of Gen. John Graves Simcoe: Commander of the "Queen's Rangers" During the Revolutionary War.* Toronto, Ont.: G. Virtue, Publisher, 1890.

Revie, Linda L. *The Niagara Companion: Explorers, Artists, and Writers at the Falls, from Discovery through the Twentieth Century.* Waterloo, Ont.: Wilfrid Laurier University Press, 2003.

Ressa, Alfonso M. *Paolo Busti: Chapter of American History, 1798–1824.* Philadelphia, PA., 1957.

Riddell, Renwick W. *The Life of William Dummer Powell, first judge at Detroit and fifth Chief Justice of Upper Canada.* Lansing, MI: Michigan Historical Commission, 1924.

Simcoe, Elizabeth, and Ross J. Robertson. *The Diary of Mrs. John Graves Simcoe.* Toronto, Ont.: William Briggs, 1911.

Scaggs, David Curtis, ed. *The Sixty Years War for the Great Lakes, 1754–1814.* East Lansing, MI: Michigan State University Press, 2012.

Seibel, George. *Bridges Over The Niagara Gorge.* Niagara Falls, Ont.: Niagara Falls Bridge Commission, 1991.

Seibel, George, *The Niagara Portage Road: 200 Years 1790–1990.* Niagara Falls, Ont.: The City of Niagara Falls Canada, 1990.

Sellers, Charles. *The Market Revolution: Jacksonian America 1815–1846.* Oxford, UK: Oxford University Press, 1991.

Severance, Frank. *An Old Frontier of France.* 2 Vols. New York: Dodd, Mead, and Company, 1917.

Severance, Frank. *Studies of the Niagara Frontier.* Buffalo, NY: Published by the Buffalo Historical Society, 1911.

Shank, William H. *Towpaths to Tugboats: A History of American Canal Engineering.* York, PA: American Canal and Transportation Center, 1982.

Shallott, Todd. *Structures in the Stream: Water, Science and the Rise of the United States Army Corp of Engineers.* Austin, TX: University of Texas Press, 1994.

Shaw, Ronald. *Erie Water West: A History of the Erie Canal, 1792–1854.* Lexington, KY: The University Press of Kentucky, 1966.

Shaw, Ronald. *Canals for a Nation: The Canal Era in the United States, 1790–1860.* Lexington, KY: The University Press of Kentucky, 1990.

Sheppard, George. *Plunder, Profit and Paroles: A Social History of the War of 1812 in Upper Canada.* Montreal and Kingston: Mc Gill-Queen's University Press, 1994.

Sheriff, Carol. *The Artificial River: The Erie Canal and the Paradox of Progress, 1817–1862.* New York: Hill and Wang, 1996.

Shortt, Edward, ed. *Perth Remembered.* Perth, Ont., Canada: Printed by Mortimer Ltd., 1967.

Smith, Perry. *History of the City of Buffalo and Erie County.* 2 Vols. New York: D. Mason & Company, Publishers, 1884.

Snodgrass, Ellen. *The Underground Railroad: an encyclopedia of people, places, and operations.* New York: M. E. Sharpe, 2008.

Snyder, Charles M. *Oswego: from Buckskins to Bustles.* Port Washington, NY: Ira J. Friedman. 1968.

Spanagel, David. *DeWitt Clinton and Amos Eaton: Geology and Power in Early New York.* Baltimore, MD: Johns Hopkins University Press, 2014.

Stagg, J.C.A. *Mr. Madison's War: Politics, Diplomacy and Warfare in the Early American Republic, 1783–1830.* Princeton, NJ: Princeton University Press, 1983.

Strand, Ginger. *Inventing Niagara Falls: Beauty, Power, and Lies.* New York: Simon and Schuster Paperbacks, 2008.

Stuart, Reginald. *United States Expansionism and British North America, 1775–1871.* Chapel Hill, NC: The University of North Carolina Press, 1988.

Styran, Roberta, and Robert R. Taylor. *Mr. Merritt's Ditch.* Erin, Ont.: A Boston Mills Press Book, 1992.

Styran, Roberta, and Robert R. Taylor, *This Great National Object: Building the Nineteenth Century Welland Canals.* Montreal & Kingston: Mc Gill-Queen's University Press, 2012.

Styran, Roberta and Robert R. Taylor, *The "Great Swival Link": Canada's Welland Canal.* Toronto, Ont.: The Champlain Society, 2001.

Switzer, Richard. *Chateaubriand's Travels in America.* Lexington, KY: The University of Kentucky Press, 1969.

Syrett, Harold C. *Papers of Alexander Hamilton.* 27 Vols. New York: Columbia University Press, 1961.

Taylor, Alan. *The Civil War of 1812: American Citizens, British Subjects, Irish Rebels, and Indian Allies.* New York: Vintage Books, 2010.

Taylor, Alan. *The Divided Ground: Indians, Settlers, and the Northern Borderland of the American Revolution*. New York: Vintage Books, 2006.

Taylor, George Rogers. *The Transportation Revolution, 1815–1860*. New York: M. E. Sharpe Inc., 1951.

Tobin, Brian and Elizabeth Hulse, eds. *The Upper Canada Gazette and its Printers, 1793–1849*. Toronto, Ont.: Ontario Legislative Library, 1993.

Tucker, Gilbert N. *The Canadian Commercial Revolution*. Toronto, Ont.: McClelland and Stewart Limited, 1964.

Tuer, J. A. *An Historical Narrative of Some Important Events in the Life of the First Church, St. Catharines, 1831–1931*. Toronto, Ont., 1931.

Vance, Michael E. *Imperial Immigrants: Scottish Settlers in the Upper Ottawa Valley, 1815–1840*. Toronto, Ont.: Dundurn Press, 2012.

Vinal, Theodore. *Niagara Portage: From Past to Present*. Buffalo, NY: Foster and Stewart, 1949.

Way, Peter. *Common Labour: Workers and the Digging of North American Canals, 1780–1860*. Cambridge, UK: Cambridge University Press, 1993.

Weingardt, Richard. *Engineering Legends: Great American Civil Engineers*. Reston, VA: American Society of Engineers, 2005.

Whitman, Roger, Scott Eberle, and David A. Gerber, eds. *The Rise and Fall of a Frontier Entrepreneur: Benjamin Rathbun, "Master Builder and Architect."* Syracuse, NY: Syracuse University Press, 1996.

Wilson, Bruce. *The Enterprises of Robert Hamilton: A Study of Wealth and Influence in Early Upper Canada, 1776–1812*. Ottawa, Canada: Carleton University Press, 1983.

Whitford, Noble. *History of the Canal System of the State of New York: Together with Brief Histories of the Canals of the United States and Canada*. 2 Vols. Albany, NY: Brandow Printing Company, 1905.

Williams, Jack. *Merritt: A Canadian Before His Time*. St. Catharines, Ont.: Stonehouse Publications, 1985.

Wood, William, ed. *Select British Documents of the Canadian War of 1812*. 3 Vols. Toronto, Ont.: Greenwood Publishing Group, 1928.

Wright, Benjamin Hall. *Origin of the Erie Canal*. Rome, NY: Sandford and Carr, Book and Job Printers, 1870.

Wyckoff, William. *The Developer's Frontier: The Making of the Western New York Landscape*. New Haven, CT: Yale University Press, 1988.

ARTICLES

Aitken, Hugh G. "Yates and McIntyre: Lottery Managers." *Journal of Economic History*, Vol. 13, No. 1 (Winter 1953), 36–57.

Augspurger, Owen B., Jr. "Buffalo's Earlier Crossroads: Samuel Wilkeson and the Seaway, Skyway, and the Thruway." *Niagara Frontier*, Vol. 3, No. 9 (Spring 1956–Winter 1957), 33–41.

Buckner, Phillip. "How Canadian Historians stopped Worrying and learned to Love the Americans!" *Acadiensis*, Vol. 25, No. 2 (Spring 1996), 117–40.

Cain, Emily, "Customs Collection—and Dutiable Goods: Lake Ontario Ports, 1801–1812," *Freshwater*, Vol. 2, No. 2 (Autumn 1987), 22–27.

Cain, Emily. "Provisioning Lake Ontario Merchant Schooners, 1809–1812." *Freshwater*, Vol. 3, No. 1 (Summer 1988), 21–25.

Careless, J. M. S., "Frontierism, Metropolitanism, and Canadian History," *Canadian Historical Review*, Vol. 35, No. 1 (March 1954), 1–21.

Carp, Roger. "The Limits of Reform: Labor and Discipline on the Erie Canal." *Journal of the Early Republic*, Vol. 10, No. 2 (Summer 1990), 191–219.

Casey, Richard C. "North County Nemesis: The Potash Rebellion and the Embargo of 1807–1809." *New York Historical Society Quarterly*, Vol. 64 (January 1980), 31–49.

Chazanof, William. "Joseph Ellicott, The Embargo, and The War of 1812." *Niagara Frontier*, Vol. 10, No. 1 (1963), 1–18.

Dunnigan, Brian L. "Portaging Niagara," *Inland Seas*, Vol. 42, No. 3 (Fall 1986), 177–222.

Ellis, David M. "Rise of the Empire State, 1790–1820" *New York History*, Vol. 56, No. 1 (January 1975), 4–27.

Geddes, George. "The Erie Canal: Origin and History of the Measures that led to its Construction." Buffalo Historical Society *Publications*, Vol. 2, (Buffalo, NY, 1880), 263–304.

Hernandon, Melvin. "A Grandiose Scheme to Navigate and Harness Niagara Falls." *New York Historical Society Quarterly*, Vol. 58 (January 1974), 177–223.

Irwin, Robert J., Jr. "William Hamilton Merritt and the First Welland Canal." *Niagara Frontier*, Vol. 11, No. 4 (Winter, 1964), 105–18.

Jasen, Patricia. "Romanticism, Modernity, and the Evolution of Tourism on the Niagara Frontier, 1790–1850." *Canadian Historical Review*, Vol. 72, No. 3 (September 1991), 283–318.

Johnson, Arthur. "The Transportation Revolution." *Ontario History*, 67, No. 4 (December 1975), 199–209.

Masters, Donald C. "Evolution of a Frontiersman." *The Manitoba Arts Review*, Vol. 2, No. 1 (Spring 1940), 54–59.

McGreevy, Patrick. "The End of America: The Beginning of Canada." *The Canadian Geographer*, Vol. 32, No. 4 (1988), 307–18.

McIlwraith, Thomas F. "Freight Capacity and Utilization of the Erie and Great Lakes Canals before 1850." *Journal of Economic History*, Vol. 36, No. 4 (December 1976), 852–75.

Nordstrom, Justin. "Public Memory and Popular Culture: The Erie Canal in the Imagination of the 1920s. *New York History*, Vol. 80, No. 4 (October 1999), 423–54.

Norton, Charles. "The Old Black Rock Ferry." Buffalo Historical Society *Publications*, Vol. 1 (January 1979), 91–112.

Officer, Lawrence, and Lawrence B. Smith. "The Canadian-American Reciprocity Treaty of 1855–1866." *The Journal of Economic History*. Vol. 28, No. 4 (December, 1968), 598–623.

Paltsits, Victor Hugo. "Judge Augustus Porter, Pioneer of Niagara Falls." *New York History*, Vol. 18, No. 2 (April 1937), 136–51.

Rapp, Marvin A. "New York's Trade on the Great Lakes, 1800–1840." *New York History*, Vol. 39, No. 1 (January, 1958), 22–33.

Rawlyk, George A. "Thomas Coltrin Keefer and the St. Lawrence-Great Lakes Commercial System." *Inland Seas*, Vol. 19 (Fall 1963), 190–94.

Rezneck, Samuel. "A Traveling School of Science on the Erie Canal in 1826." *New York History*, Vol. 40, No. 3 (July 1959), 255–69.

Seelye, John. "Rational Exaltation: The Erie Canal Celebration." *Proceedings of the American Antiquarian Society*, Vol. 94 (1984), 241–67.

Shaw, Ronald. "Michigan Influences upon the Formative Years of the Erie Canal." *Michigan History*, Vol. 37, Issue 1 (March 1953), 1–18.

Sloan, Captain James. "Adventures and Recollections of a Pioneer Trade." Buffalo Historical Society *Publications*, Vol. 5 (1902), 287–318.

Stack, Debbie. "Oswego Canal." *The Best From American Canals*. (York, PA: The American Canal and Transportation Center, No. 3 (1989), 12.

Stagg, J. C. A. "Between Black Rock and a Hard Place: Peter B. Porter's Plan for an American Invasion of Canada in 1812." *Journal of the Early Republic*, Vol. 19, No. 3 (Autumn 1999), 385–422.

Stone, Colonel William Leete. "From New York to Niagara: Journal of a tour, in part by canal, in 1829." Buffalo Historical Society *Publications*, Vol. 14 (1910), 207–64.

Stuart, Reginald. "Special Interests and National Authority in Foreign Policy: American-Provincial British Links during the Embargo and War of 1812." *Diplomatic* History, Vol. 8, Issue 4 (October, 1984), 311–28.

Williamson, Chilton. "New York's Impact on the Canadian Economy Prior to the Completion of the Erie Canal." *New York History*, Vol. 24, No. 1 (January 1943), 24–38.

Willoughby, William R. "The Inception of the Erie and the Champlain Canal Projects. *Niagara Frontier*, Vol. 3, No. 4 (Winter 1957), 105–15.

Willoughby, William R. "The Impact of the Erie Canal," *Niagara Frontier*, Vol. 3, No. 1 (Spring 1956), 38–47.

Wilkenson, Samuel. "The Life of the Keel Boatmen." Buffalo Historical Society *Publications*, Vol. 5 (1902), 179–82.

Wylie, William, "Poverty, Distress, and Disease: Labour and the Construction of the Rideau Canal, 1826–1832," *Labour/Le Travail*, Vol. 11 (Spring 1983), 7–29.

DISSERTATIONS

Carp, Roger Evan. "The Erie Canal and the Liberal Challenge to Classical Republicanism, 1785–1850." PhD diss., University of North Carolina, 1986.

Grande, Peter. "The Political Career of Peter Buell Porter, 1797–1829." PhD diss., University of Notre Dame, 1971.
Lahey, William Charles. "The Influence of David Parish on the Development of Trade and Settlement in Northern New York, 1808–1822." PhD diss., Syracuse University, 1958.
McIlwraith, Thomas. "The Logistical Geography of the Great Lakes Grain Trade, 1820–1850." PhD diss., University of Wisconsin, 1973.
Moore, David R. "Canada and the United States, 1815–1830." PhD diss., University of Chicago, 1910.
Rubin, Israel I. "New York State and the Long Embargo," PhD diss., New York University, 1961.
Watson, J. W. "The Geography of the Niagara Peninsula." PhD diss., University of Toronto, 1945.

Index

Note: Page numbers in *italics* indicate illustrations.

abolitionism, 118–19, 135, 140–41; Underground Railroad and, *24*, 140, 220n22
Adams, Israel, 54
Adams, John Quincy, 159
Aitken, Hugh, 193n76
Akron, Ohio, 110
alcohol trade, 26–27, 53, 135
alcohol use, 98, 119, 135–39, 181, 227n119, 228n125
American Revolution, 15, 26, 30; Iroquois emigrants of, 14; loyalist emigrants from, 14, 86–87, 157, 188n4; peace treaty of, 13, 57, 188n4. *See also* "late loyalists"
Ann and Jane (ship), 102–4, 107
annexation, reciprocal trade versus, 226n103
Annual Fair of Upper Canada, 22

Ball, Elizabeth, 210n34
Bannan, Theresa, 219n20
Baring, Alexander, 86
Barrett, Alfred, 91–92, 100, 101, 139, 179
Barton, Benjamin, 25, 55, 56, 58. *See also* Porter, Barton and Company
Bernstein, Peter, 198n2, 203n71

Bethel Society, 136, 137, 227n117
Bible tract societies, 136, 137, 140, 181
Biddle, Nicholas, 159
Black Rock, NY, *24*; ferry at, 22, 59–60; lock at, 103; maps of, *4*, *102*; smuggling at, 24–25, 48; Underground Railroad at, 24
Bothwell, Robert, 188n4
Boulton, Henry John, 89–90
Brant, John, 55
Brant, Joseph, 55
Brebner, John Bartlett, 28
Britton, William, 155
Brock, Isaac, 154
Bronson, Alvin, 117–18, 134
Bronson and Company, 130
Brown, John (hotelier), 156–57, *163*, 164
Buffalo, N.Y., 2–3, 76, 164, 168; Black Rock rivalry with, 70; canal opening ceremony at, 71–73, 161; Erie Canal plans and, 51, 59, 61, 62, 70, 75, 154; Holland Land Company, 25, 26, 39, 45–47; maps of, *4*, *102*; saloons of, 227n119; social reform movements and, 135; Welland Canal and, 103–10

Bukowczyk, John, 11, 225n100
burned-over district, 181
Busti, Paul, 48–49
By, John, 221n38

Cadillac, Sieur de, 29
Cain, Emily, 202n55
Canada Land Company, 76
Cartwright, Richard, 38
Casanove, Theofilus, 48
cattle and horse trade, 7, 22, 26, 27, 48, 123n25, 193n78; on Welland Canal, 95, 99, 173
Champlain Canal, 92, 120. *See also* Lake Champlain
Chapin, Cyrenius, 208n7
Chateaubriand, François-René de, 20
Chautauqua Lake, 81
cheese sales, 26, 27, 46, 193n80
Chenango Canal, 101, 227n124
Chippewa (Canadian town), 19, 27, 164; ferry at, 22; maps of, *4, 102, 144*
Chippewa (ship), 164–66
cholera, 132, 179. *See also* sickness
Clinton, De Witt, 5, 59–60, *85*, 135, 191n37; on "The Cradles," 19; death of, 84; Erie Canal proposal of, 61, 64–67; at Erie Canal's opening, 71–72, 145, 161, 162; Merritt and, 89, 181; on Mohawk River–Lake Ontario route, 51–55; as New York governor, 68–69, 75–76; on Niagara portage route, 21, 59–61; Oswego Canal and, 76, 115, 122; Porter and, 67, 72, 76–77; pseudonym of, 203n72; on Queenston ferry, 22; on smuggling, 25; Welland Canal and, 84, 89, 94; on Western Inland Lock Navigation Company, 52–55
Clowes, Samuel, 212n66

Colden, Cadwallader D., 94, 161, 205n130, 234n75
Colles, Christopher, 31
Cooper, Thomas, 49
Corn Laws (U.K.), 182
Cossack (ferry), 164
Coteau du Lac, 17
"Cradles" (portage device), 19
Creighton, Donald, 134

Davison, Gideon, 169
De Witt, Simeon, 45–46
De Witt Clinton (ship), 131
Deep Cut of Welland Canal, 99–100, *102*, 155, 168–73, 175
Dickson, Thomas, 56
Drawback Laws (U.S.), 206n137
Durham, Lord, 154, 207n150
Durham boats, 52
Dwight, Theodore, *144*, 170

Eagle Hotel, 157
earth-moving equipment, 98–99, 172
Edwards, John, 120, 238n12
Ellice, Edward, 86
Ellicott, Benjamin, 18
Ellicott, Joseph, 18, 26, 39, 45–48; on Canadian trade, 62, 64; final days of, 72; on Lake Ontario route, 66; portrait of, *47*
Ellis, David, 201n46
Embargo Act (1807), 25, 44, 46, 53–54, 61
Embree, Effingham, 34
Enys, John, 19, 21
Erie (ship), 129
Erie Canal, 1–2, 43–78, 97; Canadian supporters of, 70, 85; Canadian workers on, 70–71; Clinton's proposal for, 61, 64–67; contractor problems with, 123; enlargement

of, 180; foreign investors in, 85–86; founder of, 5; immigrants' use of, 73, 76, 154; land donations for, 72, 83–84; map of, *xviii*; opening of, 71–72, 135, 145, 161–62; Oswego Canal and, 113–16; speculator corruption and, 122; tolls on, 70, 106, 109, 132, 216n138, 216n143, 218n165; tourism at, 75, 142–46, 152, 182, 229n2; Welland Canal and, 2–3, 73–75, 88–99, 105–6; women workers on, 70–71

Errington, Jane, 27, 204n92
evangelism, 135–37, 181

ferries, 4, 22–25, 46, 155–58, 234n65; horse-driven, 25, 164, 168
Fidler, Isaac, 150
fishing industry, 26, 52, 73, 120, 151, 193n79, 220n24, 232n38
Forsyth, William, 156–58, *163*, 164, 234n65
Fort Erie, 4, 22, 24, 66, *102*
Fort Erie Friendship Trail, 182
Fort Niagara, 4, 16, 38, *144*
Fort Schlosser, 18–20, 22, 50, 56, 59
Franz, Horace, 169
Fraser, Donald, 161
free trade, 103, 182; Drawback Laws and, 206n137; Hamilton on, 27–28; Jay's Treaty and, 28; Jefferson's embargo and, 53–54; McCalla on, 214n112; Gerrit Smith on, 118, 134, 225n103; Wadsworth on, 100
Fulton, Robert, 30

Gallatin, Albert, 46, 50–51, 55, 59
Galt, John, 76
Geddes, James, 45, 53, 63, 115, 179; Oswego Canal and, 124; Welland Canal and, 92, 100, 104

Genesee Valley, 8, 26, 50, 62; aqueduct of, 154–55; canal of, 94
Ghent, Treaty of (1814), 64
Gilman, Caroline, 147
Gilpin, Joshua, 30
Glazebrook, G. P. de T., 216n143
Gooding, William, 183–84
Grand Thoroughfare, 17–18
Grande, Peter, 202n60
Granger, Gideon, 84
Great Britain (ship), 132, 143, 152–53

Haines, Charles, 69
Hall, Basil, 1–2, 139, 147, 158, 174, 214n103
Hall, Francis, 212n66
Hamilton, Alexander, 27–28
Hamilton, John, 132
Hamilton, Robert, 15, 37–40, 194n94, 197n150
Hansen, Lee, 15
Hawthorne, Nathaniel, 148, 170
Hennepin, Louis, 29
Holland Land Company, 18, 25–27, 45–48, 84
Hopkins, T. S., 27
horse races, at Niagara, 23
horse-driven ferries, 25, 164, 168
Hosack, David, 76, 145
Hovey, Alfred, 155, 210n40
Huntington, George, 51
Huskisson, William, 85–86
Hutchinson, Holmes, 124

immigrants, 71, 73, 76, 151, 153–54; Irish, 31, 100–101, 220n20; "late loyalist," 14, 25, 80, 188n2; Scottish, 35, 119–20, 220n21
inclined plane, around Niagara Falls, 30
inclined railroad, 19

Ingalls, Otis, 27
Irwin, William, 187n26, 229n2, 234n71
Izard, Ralph, 16

Jackson, Andrew, 159, 167
Jackson, John, 238
Jansen, Patricia, 229n1
Jay's Treaty (1795), 28, 35
Jefferson, Thomas: Embargo Act of, 25, 44, 46, 53–54, 61
Jerome, William, 124

Kalm, Peter, 16
Keefer, George, 83–84
Keefer, Thomas Coltrin, 179, 208n12
Kerr, Robert, 55
Kingston, Ont., 177; Erie Canal news in, 44, 51, 68–72, 180; Oswego and, 119, 127, 131–33, 152; social reforms in, 136, 139–41
Koeppel, Gerard, 198n2

La Rochefoucault–Liancourt, Duc de, 23–24, 26, 27
La Salle, Sieur de, 29
Lachine Canal, 136, 179
Lafayette, Marquis de, 159, 169
Lake Champlain, *xviii*, 31, 67, 145, 153; canal on, 92, 120
Lake Ontario–Niagara route proposal, 45–50, 59, 62–63
Landon, Fred, 86, 208n9
Lansing, William E., 211n59
Larkin, Daniel, 219n17
"late loyalists," 5, 14–15, 25, 80, 188n2
Lewiston, N.Y., 164; cattle trade of, 26; "Cradles" at, 19; ferry landing at, *158*; maps of, *4*, *144*; portage route to, 56, 59; smuggling at, 48

Literature Lotteries of New York State, 86
Lockport, N.Y.: labor riots in, 98; locks at, 77, 99–100, *102*, 168–70, *171*; maps of, *102*, *144*; Ridge Road of, 155
loyalists. *See* "late loyalists"
Lundy's Lane, Battle of, 80

Mackenzie, William Lyon, 147, 155, 165, 167, 169
Madison, James, 54
manifest destiny, 204n92
Mansion House (hotel), 164
Martha Ogden (ship), 152, 164
Maude, John, 20, 37
McCalla, Douglas, 8, 193n76, 214n112
McDonnell, Alexander Yates, 211n59
McGreevy, Patrick, 3–4, 229n1
McGuire, Peter, 100–101
McKinsey, Elizabeth, 3, 159, 191n47, 236n100
McNall, Neil Adams, 8, 25
Melish, John, 32–33; on canal proposal, 49–50; on smuggling, 24, 25
Merritt, William Hamilton, 22–23, 56, 80–94, 206n145; as abolitionist, 140–41; Clinton and, 89, 181; on cross-border trade, 134; on international trade, 111–12; letters of, 208n10; Oswego Canal and, 115, 128–29; Porter and, 103; portrait of, *81*; railroad interests of, 23, 146, 181–82; during War of 1812, 80–81, 87; Welland Canal and, 80–94, 103–5; writings of, 101, 228n139; Yates and, 179
Michigan spectacle, 11, 155–68, *163*, 175, 236n100

Miller, Nathan, 68
Mohawk Indians, 55
Montreal, 16; New York City's rivalry with, 2–6, 43–44, 61–63, 68, 113; New York State trade and, 79, 89, 90, 107, 162; War of 1812 and, 24, 64
Murray, Charles Augustus, 230n13

Native Americans, 4, 55, 137, *144*
New York City, 2, 34, 64; Erie Canal's completion and, 72; merchants of, 67, 68, 71; Montreal rivalry with, 2–6, 43–44, 61–63, 68, 113; War of 1812 and, 6
New York State Barge Canal. *See* Erie Canal
Niagara (Canadian town), 164; ferry at, 22; *Michigan* spectacle at, 175, 236n100; trade of, 27, 66
Niagara Canal Company, 34–37, 46, 59, 77
Niagara Falls: Hennepin on, 29; maps of, 4, *102*; McKinsey on, 3, 229n1; *Michigan* spectacle at, 11, 155–68, *163*; suspension bridge over, 182; tourism at, 158–60, 164, 168, 170, 173–75, 183
Niagara Genesee Land Company, 56
Niagara portage, 46
Niagara portage road, 18–20, 38
Niles, Hezekiah, 77, 102, 110; on *Michigan* spectacle, 160, 165, 167
Northern Inland Lock Navigation Company, 31
Northern Tour, 10–11, 59, 144–45, 151–52, *156*. *See also* tourism

Ohio (ship), 131
Ohio Canal, 110, 124, 130
Ontario (ship), 152, 164

Ontario House (hotel), 157, 164
Oswego, N.Y., 16, 63; artist's view of, *115*; Kingston and, 119, 127, 131–33, 152; lighthouse of, 151; shipbuilding at, 131, 177; smuggling at, 54; Underground Railroad at, 220n22
Oswego Canal, 88, 113–42; Canadian workers on, 71, 101; Clinton and, 76, 115, 122; construction problems on, 122–26; Erie Canal and, 113–16, 121; Basil Hall on, 1; immigrants' use of, 131, 154; importance of, 6–7, 51–53, 127–28, 132–34, 141–42, 186n14; map of, xviii; Merritt and, 76, 115, 128–29; opening of, 126; Rideau Canal and, 123; Scottish workers on, 119–20, 220n21; sections of, 122; tourism at, *115*, 142, 144, 145, 148–52, 182; traffic on, 219n17; Welland Canal and, 93, 105, 111–12, 127–34
Oswego Canal Company, 117–18, 120–21, 124
Oswego River, 16, 31, 117, 186n18
Oswego–Lake Ontario channel, 16

Panic of 1837, 2, 181
Paris, Treaty of (1783), 13, 57, 188n4
Parish, David, 49
Patch, Sam, 166, 236n101
Pavillion Hotel, 157, 164
Peck, S., 123
Phelps, Charlotte, 97–98, 138
Phelps, Oliver, 94, 97–99, 101, 122, 125, 179; as abolitionist, 140–41; social reform movements and, 136–39; on Welland Deep Cut, 172
Phelps, Orson, 94, 179
Pickering, Joseph, 172–73

Port Colborne (Canada), *102*, 105
Port Dalhousie (Canada), *102*, 105
Porteous, John, 39
Porter, Augustus, 19, 56, 165, 233n64; business interests of, 60, 77, 157–58, 201n55; Oswego Canal and, 63, 116
Porter, Barton and Company, 55, 58, 59, 161; Oswego Canal and, 118; portage business of, 25, 56
Porter, Peter B., 19, 51, 55–58, 86, 165; business interests of, 60, 77, 201n55; Clinton and, 67, 72, 76–77; on cross-border relations, 58, 202n62; Erie Canal proposals and, 69–70; Merritt and, 103; as military hero, *57*, 64; Oswego Canal and, 63, 115, 116; portrait of, *57*; as Secretary of War, 159; tourism development by, 60, 77, 202n55; Welland Canal and, 103
potash, 27, 48, 52, 53, 61, 62, 107, 110
Powell, Anne, 16–17, 19, 20
Powell, William (of Canada), 73–74
Powell, William Dummer (of U.S.), 189n17
Prendergast, Catharine, 80, 87, 208n10
Prendergast, J., 80, 81, 208n11
Prescott, Benjamin, 34, 35
Pulteney estate, 35

Queenston (Canada), 54–55, 164; ferry at, 22, *158*; map of, *4*

R. H. Boughton (ship), 102, 103, 107
railroads, 19, 146, 181–82
Rattlesnake (ship), 132
Reciprocity Treaty (1854), 182
Revie, Linda, 229n1

Rideau Canal, 130, 221n38; Oswego Canal and, 123; Scottish workers on, 119–20; workers' wages of, 221n40
Roberts, Nathan, 77, 168–69, 171, 179
Rough, James, 161, 166
Russell, Peter, 28

Sabbath-breaking laws, 135–36
Sage, Nathan, 54
Salina, N.Y., 117, *149*, 149–51
Salina Canal, 116
salmon fishing, 26, 73, 193n79, 220n24, 232n38. *See also* fishing industry
salt, 52–54, 82, 110, 130, *144*; from Salina, 116, *149*, 149–51; from Syracuse, 126–27, 148
Schultz, Christian, 22
Seamen's Friend Society, 136, 137, 139–40
Sedgwick, Catharine, 146–47
Seneca Chief (ship), 71–72, 161, 162
Shaw, Ronald, 58, 62, 186n18, 197n2; on Erie Canal, 72, 206n145; on Welland Canal, 92
Sheriff, Carol, 220n23, 226n107, 227n117
sickness: near Welland Canal, 137; on New York canals, 116, 125, 132, 138, 179, 222n47
Simcoe, John Graves, 15, 21, 22
Sloan, James, 214n114
Smith, Gerrit, *118*, 118–19; as abolitionist, 118, 140; on free trade, 118, 134, 225n103; on opening of Oswego Canal, 129–30; as temperance reformer, 140
Smith, William, of Waterloo, 74
smuggling, 23–25, 46, 53, 61, 102, 157; at Black Rock, 24–25, 48;

Hamilton on, 27–28; hired agents for, 7–8, 54; at Lewiston, 48; at Oswego, 54; of tea, 24, 154, 233n57
Smyth, Charles, 133–34
Snyder, Charles, 114
social reform movements, 135–39, 178; abolitionism and, *118*, 135, 140; temperance and, 98, 135–39, 181
Split Rock quarry, 100–101
St. Catherines (Canada), 82; escaped slave settlement in, 140–41; evangelicalism in, 136–37; map of, *102*; Welland Canal and, 75, 93, 109
stagecoaches, 145–46, *156*, 164
Steele, Orlo, 118
Stephen Girard (ship), 111
Strand, Ginger, 229n1
Street, Samuel, 66
Stuart, James, 148, 154
Stuart, Reginald, 15, 26; on cross-border economy, 204n92; on Oswego flour mills, 225n94
Swallow (ship), 72–73

Tatham, William, 30
Taylor, Alan, 202n62
Taylor, George Rogers, 186n13
tea, 39, 53, 134; smuggling of, 24, 154, 233n57
temperance movement, 98, 135–39, 181
Thomas, David, 91, 124
Tibbett(s), Hiram, 83
tolls: on Erie Canal, 70, 106, 109, 132, 216n138, 216n143, 218n165; as trade impediment, 22, 68; on Welland Canal Company, 106, 132, 211n50

Tompkins, Daniel, 67
Toronto (York), 54, 108
tourism, 3–4, 8–11, 20, 143–75; at Coteau du Lac, 17; at Erie Canal, 75, 142–46, 152, 182, 229n2; at Lewiston, 19; at Lockport Locks, 168–70; at Niagara Falls, 158–60, 164, 168, 170, 173–75, 183; at Oswego Canal, *115*, 142, 144, 145, 148–52, 182; Porter brothers' development of, 60, 77, 202n55; War of 1812 and, 145, 154, 168; at Welland Deep Cut, 168–73, 175. *See also* Northern Tour
Troup, Robert, 36
Tubman, Harriet, 140–41
Tucker, Gilbert, 73, 206n137, 211n50

Underground Railroad, *24*, 140, 220n22. *See also* abolitionism
United States (ship), 152–53

Van Rensselaer, Stephen, 56, 59
Vandalia (ship), 177, *178*
Victory (ship), 130–31
violence, among canal workers, 111, 136, 138, 214n102

Wadsworth, James, 53
Walk-in-the-Water (ship), 65
Wallenda, Nik, 183
Walton and Willet Company, 131
War of 1812, 5, 44, 63–64, 183–84; borderland before, 13–41; embargoes during, 54; Erie Canal and, 61; tourism and, 145, 154, 168
Watson, Elkanah, 33–34, 36, 37
Watson, James, 34, 35
Way, Peter, 123, 214n102, 221n40
Wedding of the Waters, 71–72, 76
Weld, Isaac, 25, 40–41

Welland (Canada), *102*
Welland Canal, 6, 79–112, 179–81;
  Boulton on, 90; "Celebration Song"
  of, 107; Clinton and, 84, 89, 94;
  contractor problems with, 123,
  171–73; Deep Cut of, 99–100, *102*,
  155, 168–73, 175; dimensions of,
  93; Erie Canal and, 2–3, 73–75,
  88–99, 105–10; Fort Erie Friendship
  Trail and, 182; Basil Hall on, 1–2;
  importance of, 90–91, 93, 105–12,
  142, 149; labor unrest and, 98;
  landslides on, 99–100; maps of,
  xviii, *102*, *144*; Merritt and, 80–96,
  101, 103–5; Native Americans
  near, 137; New York investors
  in, 77, 93; opening of, 102–5,
  107, 129, 207n2; Oswego Canal
  and, 93, 105, 111–12, 127–34;
  proposal for, 82–83; quarry of,
  100–101; social reform movements
  and, 135; temperance reforms and,
  138; tourism and, 144, 168–73,
  175; U.S. workers on, 91–101,
  95; violence on, 111, 136, 138,
  214n102; women stockholders of, 86
Welland Canal Company, 55, 83–112,
  179–80; British investors in,
  85–86; creation of, 56, 83; housing
  provided by, 97–98; tolls on, 106,
  132, 211n50; U.S. investors in, 41,
  85–88, 211n59; wages paid by, *95*,
  96–97, 123
Welland House (hotel), 128, 129,
  148–49

Wellington, Duke of, 85
Western Canal Company, 61
Western Inland Lock Navigation
  Company, 31–33, 36, 39, 70;
  Clinton on, 52–55; De Witt and,
  45; Gallatin on, 51; Holland Land
  Company and, 48
Western Seamen's Friend Society, 136
Weston, William, 31
Whitford, Noble, 196n126, 219n6
Whitney, Joseph, 153
Whitney, Parkhurst, 156–58, *163*
Wilker, J., 66
Wilkeson, Samuel, 94
Williams, John, 34
Williamson, Charles, 34, 35, 46
Willoughby, William R., 198n2
Winnebago (ship), 130, 137
Winslow, Gordon, 137
women, 10, 98; as Erie Canal
  workers, 70; as social reformers,
  139, 188n27; as Welland Canal
  stockholders, 86
Wood, Eleazer, 145
Wright, Benjamin, 45, 92, 179,
  198n5; Oswego Canal plans of, 122,
  124

Yates, John B., 86–88, 106, 109,
  211n59; Merritt and, 179; Oswego
  Canal and, 130; Polytechnic
  Institute founded by, 141; portrait
  of, *87*
York (Toronto), 54, 108
Young Lion of the West (ship), 161